SMOKELESS TOBACCO PRODUCTS

Emerging Issues in Analytical Chemistry

Series Editor
Brian F. Thomas

Co-published by Elsevier and RTI Press, the *Emerging Issues in Analytical Chemistry* series highlights contemporary challenges in health, environmental, and forensic sciences being addressed by novel analytical chemistry approaches, methods, or instrumentation. Each volume is available as an e-book, on Elsevier's ScienceDirect, and via print. Series editor Dr. Brian F. Thomas continuously identifies volume authors and topics; areas of current interest include identification of tobacco product content prompted by regulations of the Family Tobacco Control Act, constituents and use characteristics of e-cigarettes and vaporizers, analysis of the synthetic cannabinoids and cathinones proliferating on the illicit market, medication compliance and prescription pain killer use and diversion, and environmental exposure to chemicals such as phthalates, endocrine disrupters, and flame retardants. Novel analytical methods and approaches are also highlighted, such as ultraperformance convergence chromatography, ion mobility, in silico chemoinformatics, and metallomics. By highlighting analytical innovations and new information, this series furthers our understanding of chemicals, exposures, and societal consequences.

2016 *The Analytical Chemistry of Cannabis: Quality Assessment, Assurance, and Regulation of Medicinal Marijuana and Cannabinoid Preparations*
Brian F. Thomas and Mahmoud ElSohly

2016 *Exercise, Sport, and Bioanalytical Chemistry: Principles and Practice*
Anthony C. Hackney

2016 *Analytical Chemistry for Assessing Medication Adherence*
Sangeeta Tanna and Graham Lawson

2017 *Sustainable Shale Oil and Gas: Analytical Chemistry, Geochemistry, and Biochemistry Methods*
Vikram Rao and Rob Knight

2017 *Analytical Assessment of e-Cigarettes: From Contents to Chemical and Particle Exposure Profiles*
Konstantinos E. Farsalinos, I. Gene Gillman, Stephen S. Hecht, Riccardo Polosa, and Jonathan Thornburg

2018 *Doping, Performance-Enhancing Drugs, and Hormones in Sport: Mechanisms of Action and Methods of Detection*
Anthony C. Hackney

2018 *Inhaled Pharmaceutical Product Development Perspectives: Challenges and Opportunities*
Anthony J. Hickey

2018 *Detection of Drugs and Their Metabolites in Oral Fluid*
Robert M. White, Sr., and Christine M. Moore

2020 *Analytical Methods for Biomass Characterization and Conversion*
David C. Dayton and Thomas D. Foust

SMOKELESS TOBACCO PRODUCTS

Characteristics, Usage, Health Effects, and Regulatory Implications

Edited by

WALLACE B. PICKWORTH, PHD
Research Leader
Battelle Memorial Institute
Baltimore, MD, United States

Elsevier
Radarweg 29, PO Box 211, 1000 AE Amsterdam, Netherlands
The Boulevard, Langford Lane, Kidlington, Oxford OX5 1GB, United Kingdom
50 Hampshire Street, 5th Floor, Cambridge, MA 02139, United States

Published in cooperation with RTI Press at RTI International, an independent, nonprofit research institute that provides research, development, and technical services to government and commercial clients worldwide (www.rti.org). RTI Press is RTI's open-access, peer-reviewed publishing channel. RTI International is a trade name of Research Triangle Institute.

Library of Congress Cataloging-in-Publication Data
A catalog record for this book is available from the Library of Congress

British Library Cataloguing-in-Publication Data
A catalogue record for this book is available from the British Library

ISBN: 978-0-12-818158-4

For information on all Elsevier publications visit our website at
https://www.elsevier.com/books-and-journals

Publisher: Susan Dennis
Acquisition Editor: Kathryn Eryilmaz
Editorial Project Manager: Laura Okidi
Production Project Manager: Surya Narayanan Jayachandran
Cover Designer: Matthew Limbert
Cover Art: Dayle G. Johnson

Typeset by TNQ Technologies

Dedication

This book is dedicated to my granddaughters Adalynn Mala and Camille Simmons Dosi, and to all of the children in their generation, with the hope that the advancements in health science will reduce the scourge of tobacco-related disease.

Contents

Contributors

Laura Akers
Oregon Research Institute, Eugene, OR, United States

Olalekan A. Ayo-Yusuf
Sefako Makgatho Health Sciences University Medunsa, Pretoria, South Africa

Kevin P. Conway
RTI International, Research Triangle Park, NC, United States

Karl O. Fagerström
Fagerström Consulting, Vaxholm, Sweden

Judith S. Gordon
University of Arizona, Tucson, AZ, United States

Dorothy K. Hatsukami
University of Minnesota, Minneapolis, MN, United States

Stephen S. Hecht
University of Minnesota, Minneapolis, MN, United States

Jack E. Henningfield
Pinney Associates, Bethesda, MD and The Johns Hopkins University School of Medicine, Baltimore, MD, United States

Lynn C. Hull
U.S. Food and Drug Administration, Silver Spring, MD, United States

Dalia Khoury
RTI International, Research Triangle Park, NC, United States

Bartosz Koszowski
Battelle Memorial Institute, Baltimore, MD, United States

Gretchen McHenry
RTI International, Research Triangle Park, NC, United States

Steven E. Meredith
U.S. Food and Drug Administration, Silver Spring, MD, United States

Devon Noonan
Duke University, Durham, NC, United States

Richard J. O'Connor
Roswell Park Comprehensive Cancer Center, Buffalo, NY, United States

Mark J. Parascandola
National Cancer Institute, Bethesda, MD, United States

Stephanie J. Parker
RTI International, Research Triangle Park, NC, United States

Wallace B. Pickworth
Battelle Memorial Institute, Baltimore, MD, United States

Lars M. Ramström
Institute for Tobacco Studies, Täby, Sweden

Vaughan W. Rees
Harvard T.H. Chan School of Public Health, Boston, MA, United States

Herbert H. Severson
Oregon Research Institute, Eugene, OR, United States

Stephen B. Stanfill
U.S. Centers for Disease Control and Prevention, Atlanta, GA, United States

Irina Stepanov
University of Minnesota, Minneapolis, MN, United States

Foreword

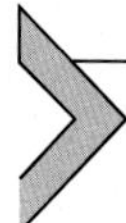

A place for smokeless tobacco in public health with science-guided tobacco regulation?

Smokeless tobacco, "you've come a long way, baby," to echo the phrase from the 1968 Virginia Slims cigarette advertisement. From a mainly older man's habit in the 1960s in the United States with products and spitting behavior that seemed disgusting to nonusers, to the emergence of starter products designed for youth in the United States in the 1970s that turned the demographics of US users upside down,[1,2] and more recently to substantially cleaned up and more population-acceptable harm reduction products in the 21st century. As evidenced in this volume, smokeless tobacco has come a long way indeed.

There were two landmarks in the cleaning up of smokeless tobacco, and it began with Swedish snus. In 1976 the first popular portion-packed (spit-free) snus product was launched, and in 1981 there was a switch to a modernized manufacturing method. The established non-fermentation heat treatment character of the method was retained, and various quality improvements of the heat treatment process were introduced, including a completely closed manufacturing process to minimize contamination. These efforts substantially reduced nitrosamines and other toxicant levels. Swedish snus is regulated under the Swedish Food Act, and Swedish health authorities encouraged improvement of snus quality. In addition, major manufacturers developed product standards to ensure low levels of toxicants (i.e, the GothiaTek Standard[3]). See additional discussion in chapters by Stanfill, Hecht, and Stepanov and Hatsukami in this volume.[4–6]

At the time of the initial drafting of this foreword, three low toxicant smokeless products were under consideration for potential official US Food and Drug Administration (FDA) designation as Modified Risk Tobacco Products (MRTPs), namely, Swedish Match's General Snus, Reynolds American's Camel Snus, and US Smokeless Tobacco Company's Copenhagen Snuff Fine Cut.[7] The Swedish Match product was the first tobacco product to have its Premarket Tobacco Product Application (PMTA) approved by FDA, and it was approved with a statement from FDA that would have seemed unimaginable to most tobacco control experts just a decade earlier: "[t]he PMTA decisions for these products reflect evidence showing that these products, marketed as described in the manufacturer's application, would result in a low likelihood of new initiation, delayed cessation or relapse... [and] that these products would likely provide less toxic options if current adult smokeless tobacco users used them exclusively."[8]

During the finalization of this foreword, an important event in the history of tobacco product regulation was announced by FDA under the following headline on their website: "FDA grants first-ever modified risk orders to eight smokeless tobacco products. FDA concludes completely switching from cigarettes to these authorized products lowers certain health risks." FDA's MRTP authorization was for eight Swedish Match snus products sold under the "General" brand name.

A harbinger of FDA's MRTP authorization and increased tobacco control community acceptance of smokeless tobacco as a harm reduction approach was a report from the Royal College of Physicians of London[9] that endorsed the use of smokeless tobacco in place of cigarettes for people who could not give up tobacco use; it discussed Swedish snus as an exemplary potential harm reduction tobacco product. This was followed by a report from the US-based Strategic Dialogue, chaired by Dorothy Hatsukami and Mitch Zeller and which included a diverse array of tobacco control experts.[10] Without endorsing smokeless tobacco per se as a harm reduction option for cigarette smokers, their report made clear that combustible tobacco products accounted for most tobacco-related morbidity and mortality and that public health needed to consider noncombustible options for people who are unable to completely give up nicotine.[10] The landmark 50th anniversary Surgeon General's Report on Tobacco and Health took a generally similar position.[11]

Research published since the Swedish Match PMTA was filed with FDA in 2014 provides further evidence that snus use both hampers initiation of smoking and encourages and facilitates cessation.[12,13] Whereas it is clear that oral smokeless tobacco does not cause lung diseases, there have been studies suggesting a possible link between snus and certain cancers, cardiovascular diseases, and type 2 diabetes. However, the most recent research suggests that these risks, if real, are substantially lower as compared to cigarette smoking.[14–19] See additional discussion in the chapter by McHenry et al. in this volume.[20]

Nonetheless, in the United States and many other countries, there remains resistance to the acceptance of smokeless tobacco as part of a comprehensive public strategy to reduce combusted tobacco smoke exposure, in part due to the legacy of aggressive youth-targeted product development and marketing that emerged in the 1970s and 1980s.[1,2] See additional discussion of past and present smokeless tobacco marketing by Rees, Ayo-Yusuf, and O'Connor in this volume.[21] A further complication in acceptance of smokeless tobacco as a form of tobacco harm reduction is that it is a diverse product category, with many products in India, North Africa, and Southeast Asia, in particular, that have very high levels of toxicants and likely carry a higher risk of disease than lower toxicant products, as discussed by Stanfill in this volume.[22] See additional discussion by Parascandola and Pickworth.[7]

Among public health leaders who do see smokeless tobacco as a potential asset in reducing smoking-associated morbidity and mortality, there is widespread agreement that such tobacco harm reduction efforts would be ideally developed and overseen in a science-guided regulated environment.[7,10,23–27]

At least three important advances have opened the door to broader acceptance of smokeless tobacco as a form of harm reduction for people unable or unwilling to completely give up tobacco. First, as illustrated by this volume, is the advancement of science that can contribute to tobacco- and health-related policy and regulation, thereby providing a regulatory framework for labeling and other messaging that could appropriately communicate this information. Second is the increasing recognition that it is combusted products in general and cigarettes in particular that overwhelming contribute to tobacco-related morbidity and mortality, and that on current course annual smoking-caused deaths rates will continue to increase for decades to come.[23,27–29] Third is what is often referred to as the Swedish Experiment, which showed that large-scale substitution of noncombusted forms of nicotine (primarily in the form of snus but also by way of nicotine replacement medicines) produced the first clear reversal of smoking-associated morbidity in the world at the population level, with a leveling off of smoking-related death rate by about the year 2000 and subsequent steady decline.[12–17] This has resulted in Sweden achieving the lowest tobacco-related deaths in the European Union.[30,31]

The rationale and approach in Sweden have been documented elsewhere, including in a commentary by Henningfield and Fagerström in 2001.[32] In brief, the Swedish Experiment began in the 1960s following the reports by the 1962 Royal College of Physicians of London[33] and the US Surgeon General[34] that cigarette smoking caused cancer and other diseases. The concept advocated by Swedish health experts was to augment efforts to prevent smoking and support cessation with acceptance and eventually encouragement of the use of smokeless tobacco in place of cigarettes.

At the same time in Sweden, cigarette companies began to market so-called light/low tar cigarettes, as they did globally. Swedish women quickly adopted "light" cigarette brands. It was decades before it became evident that such cigarettes did not reduce morbidity and mortality risks.[35] In contrast, men were more likely to transfer their tobacco use from cigarettes to snus, and their risk of lung cancer and other smoking-related diseases began to decline.[30,31] The leading Swedish tobacco company, Swedish Match, began to develop progressively lower toxin smokeless tobacco products, and in the late 1990s discontinued marketing cigarettes in Sweden. Overall nicotine consumption and prevalence of nicotine product use did not substantially decline, but in men the source of nicotine had shifted substantially from primarily cigarettes to snus. By 2001, Henningfield and Fagerström described it as a promising experiment in progress with evidence emerging that smoking-associated disease was beginning to decline.

Whereas the Swedish Experiment remains in progress, by the early 2000s it became increasingly accepted that smokeless tobacco was less harmful than cigarettes, and that consideration should be given to how policy and regulation might support transition away from cigarettes for people unable to give up tobacco and nicotine completely, but without undermining smoking prevention and tobacco use cessation efforts.[10,11,36–41] In the United States, the regulatory pathway for such a potential harm reduction use of oral smokeless tobacco and other products is included in the 2009 Family Smoking Prevention and Tobacco Control Act that established the Center for Tobacco Products within the FDA and included a mechanism to approve potential MRTPs.[26]

The Swedish Experiment remains ongoing with a steady output of data, and main findings seem clear: namely, even without reduction in nicotine use, morbidity and mortality can be substantially reduced by substitution of low toxicant smokeless tobacco for cigarettes. Indeed, the experience from Sweden means that, among males, 50 years of substitution of snus for cigarettes has been the major contributor to achieving the lowest prevalence of smoking and the lowest level of mortality attributable to tobacco in the Western World.[35,36]

Big questions remain, however, as discussed by Parascandola and Pickworth and in other chapters in this volume. They include how to increasingly help cigarette smokers who are unable to completely give up tobacco or nicotine to reduce their exposure to tobacco smoke and its associated morbidity and mortality. For many people conventional treatment modalities will suffice. For others, as stated by Gottlieb and Zeller, "the availability of potentially less harmful tobacco products could reduce risk while delivering satisfying levels of nicotine for adults who still need or want it."[39] For an appreciation of how far the science and potential products have advanced since the 1990s, see discussions over the past two decades about how such harm reduction approaches could complement prevention and treatment efforts to accelerate progress in reducing tobacco-associated morbidity and mortality by Warner, Slade, Sweanor, and others [42–44] as well as in the 2014 US Surgeon General's report.[11] The science to date, as highlighted by much of the science presented in this volume, suggests harm reduction approaches including appropriately regulated low toxicant smokeless tobacco are an achievable strategy.

Continued research to advance knowledge and guide regulation and policy is critical to continuing progress. We are pleased to welcome this volume and look forward to an update, hopefully within a few years, describing further advances in nicotine and tobacco science as well as science-guided tobacco control efforts, including tobacco product regulation.

Jack E. Henningfield, PhD
Vice President, Research, Health Policy, and Abuse Liability,
Pinney Associates, Bethesda, Maryland, United States
and
Professor, Adjunct, Behavioral Biology, Department of Psychiatry
and Behavioral Sciences, The Johns Hopkins University School of
Medicine, Baltimore, Maryland, United States
Lars M. Ramström, PhD
Principal Investigator, Institute for Tobacco Studies, Täby, Sweden
Karl O. Fagerström, PhD
Fagerström Consulting, Vaxholm, Sweden

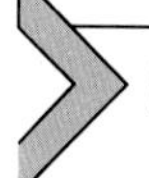

Disclosures

From 2015 to October 2019, Dr. Henningfield provided scientific and regulatory consulting services through Pinney Associates to Niconovum USA, Inc., R.J. Reynolds Vapor Company, and RAI Service Company, all subsidiaries of Reynolds American Inc. (now owned by British American Tobacco) on smoking cessation and harm reduction, explicitly excluding consultation on combusting cigarettes. Since October 2019, he has consulted on advancing relative risk-based regulation of nicotine and tobacco products through Pinney Associates to JUUL Labs, Inc. Also through Pinney Associates, he consults to pharmaceutical companies on the development and regulation of new medicines and formulations for pain, addiction, epilepsy, and other central nervous system disorders, and to the dietary supplement industry in support of central nervous system active dietary ingredients and cannabis-derived products.

Dr. Ramström is leader of the Institute for Tobacco Studies, Sweden, an independent research institution working in international collaboration with scientists dealing with epidemiological research on tobacco matters. He has received consulting fees from pharmacological companies that develop and market products for smoking cessation.

Dr. Fagerström has received consulting and speaking fees from many companies that develop or market pharmacological and behavioral treatments for smoking cessation. He currently receives consulting fees from Swedish Match and has received fees in the past from Philip Morris International and BAT to assist their efforts to develop less-risky tobacco products.

References

1. U.S. Department of Health and Human Services. *The Health Consequences of Using Smokeless Tobacco*. A Report to the Advisory Committee to the Surgeon General. Bethesda, MD: U.S. Department of Health and Human Services, Public Health Service; 1986. NIH Pub. No. 86-2874. https://profiles.nlm.nih.gov/spotlight/nn/catalog/nlm:nlmuid-101584932X65-doc. Accessed October 8, 2019.
2. U.S. Department of Health and Human Services. *Preventing Tobacco Use Among Young People: A Report of the Surgeon General (link is external)*. Atlanta, GA: U.S. Department of Health and Human Services, Public Health Service, Centers for Disease Control and Prevention, National Center for Chronic Disease Prevention and Health Promotion, Office on Smoking and Health; 1994.
3. Rutqvist LE, Curvall M, Hassler T, Ringberger T, Wahlberg I. Swedish snus and the GothiaTek® standard. *Harm Reduct J*. 2011;8:11.
4. Stanfill SB, Chapter 8, this volume.
5. Hecht SS, Chapter 7, this volume.
6. Stepanova I, Hatsukami DK, Chapter 6, this volume.
7. Parascandola MJ, Pickworth WB, Chapter 9, this volume.
8. Clissold DB. *Swedish Match North America, Inc. First to Cross FDA's PMTA Finish Line*. FDA Law Blog; 2015. http://www.fdalawblog.net/2015/11/swedish-match-north-america-inc-first-to-cross-fdas-pmta-finish-line/. Accessed October 4, 2015.
9. Royal College of Physicians. Harm reduction in nicotine addiction: helping people who can't quit. In: *A Report by the Tobacco Advisory Group of the Royal College of Physicians*. London: RCP; 2007.
10. Zeller M, Hatsukami D, Backinger C, et al. The strategic dialogue on tobacco harm reduction: a vision and blueprint for action in the United States. *Tob Control*. 2009;18:324–332.

11. U.S. Department of Health and Human Services. *The Health Consequences of Smoking—50 Years of Progress: A Report of the Surgeon General.* Atlanta, GA: US Department of Health and Human Services, Centers for Disease Control and Prevention, National Center for Chronic Disease Prevention and Health Promotion, Office on Smoking and Health; 2014. https://www.ncbi.nlm.nih.gov/books/NBK179276/pdf/Bookshelf_NBK179276.pdf. Accessed October 8, 2019.
12. Ramstrom L, Borland R, Wikmans T. Patterns of smoking and snus use in Sweden: implications for public health. *Int J Environ Res Public Health.* 2016;13:1110.
13. Lund I, Lund KE. How has the availability of snus influenced cigarette smoking in Norway? *Int J Environ Res Public Health.* 2014;11(11):11705—11717.
14. Araghi M, Rosaria Galanti M, Lundberg M, et al. Use of moist oral snuff (snus) and pancreatic cancer: pooled analysis of nine prospective observational studies. *Int J Cancer.* 2017;141(4):687—693.
15. Asthana S, Labani S, Kailash U, Sinha DN, Mehrotra R. Association of smokeless tobacco use and oral cancer: a systematic global review and meta-analysis. *Nicotine Tob Res.* 2019;21(9):1—10.
16. Hansson J, Galanti MR, Hergens MP, et al. Use of snus and acute myocardial infaction: pooled analysis of eight prospective observational studies. *Eur J Epidemiol.* 2012;27:771—779.
17. Pemberton MN. Oral cancer and tobacco: developments in harm reduction [published online ahead of print November 2, 2018] *Br Dent J.* 2018. https://doi.org/10.1038/sj.bdj.2018.928.
18. Rasouli B, Andersson T, Carlsson PO, et al. Use of Swedish smokeless tobacco (snus) and the risk of Type 2 diabetes and latent autoimmune diabetes of adulthood (LADA). *Diabet Med.* 2017;34(4):514—521.
19. Rostron BL, Chang JT, Anic GM, Tanwar M, Chang CM, Corey CG. Smokeless tobacco use and circulatory disease risk: a systematic review and meta-analysis. *Open Heart.* 2018;5(2):e000846.
20. McHenry G, Parker SJ, Khoury D, Conway KP, Chapter 3, this volume.
21. Rees VW, Ayo-Yusuf OA, O'Connor RJ, Chapter 2, this volume.
22. Stanfill SB, Connolly GN, Zhang L, et al. Global surveillance of oral tobacco products: total nicotine, unionised nicotine and tobacco-specific N-nitrosamines. *Tob Control.* 2011;20(3):e2.
23. Slade J, Henningfield JE. Tobacco product regulation: context and issues. *Food Drug Law J.* 1998;53(suppl):43—74.
24. Royal College of Physicians of London. *Protecting Smokers, Saving Lives: The Case for a Tobacco and Nicotine Regulatory Authority.* London: Royal College of Physicians; 2002.
25. Bates C, Fagerström K, Jarvis MJ, Kunze M, McNeill A, Ramström L. European union policy on smokeless tobacco: a statement in favour of evidence based regulation for public health. *Tob Control.* 2003;12(4):360—367.
26. U.S. Congress. Family Smoking Prevention and Tobacco Control Act. Public Law # 111-31. 2009.
27. U.S. Department of Health Human Services. *The Health Consequences of Smoking—50 Years of Progress: A Report of the Surgeon General, 2014.* Atlanta, GA: U.S. Department of Health Human Services, Centers for Disease Control and Prevention, National Center for Chronic Disease Prevention and Health Promotion, Office on Smoking and Health; 2014.
28. Henningfield JE, Slade J. Tobacco-dependence medications: public health and regulatory issues. *Food Drug Law J.* 1998;53(suppl):75—114.
29. World Health Organization. *Advancing Knowledge on Regulating Tobacco Products (Monograph).* Geneva: World Health Organization; 2001. https://www.who.int/tobacco/publications/prod_regulation/OsloMonograph.pdf?ua=1. Accessed October 8, 2019.

30. Myrdal G, Lambe M, Bergström R, Ekbom A, Wagenius G, Ståhle E. Trends in lung cancer incidence in Sweden with special reference to period and birth cohorts. *Cancer Causes Control.* 2001;12(6):539–549.
31. Ramström L, Wikmans T. Mortality attributable to tobacco among men in Sweden and other European countries: an analysis of data in a WHO report. *Tob Induc Dis.* 2014;12(1):14.
32. Henningfield JE, Fagerstrom KO. Swedish match company, Swedish snus and public health: a harm reduction experiment in progress? *Tob Control.* 2001;10(3):253–257.
33. Royal College of Physicians of London. Smoking and health. *A Report on Smoking in Relation to Lung Cancer and other Diseases.* London: Royal College of Physicians; 1962.
34. United States Public Health Service. *Smoking and Health: Report of the Advisory Committee to the Surgeon General of the Public Health Service.* Washington, DC: US Department of Health, Education, and Welfare; 1964.
35. U.S. Department of Health and Human Services, National Institutes of Health, National Cancer Institute. Monograph 13: risks associated with smoking cigarettes with low machine-measured yields of tar and nicotine. In: *National Cancer Institute's (NCI) Smoking and Tobacco Control Program Monograph Series*; 2001. https://cancercontrol.cancer.gov/brp/tcrb/monographs/13/m13_complete.pdf. Accessed October 8, 2019. Updated March 1, 2012.
36. Institute of Medicine. *Clearing the Smoke: Assessing the Science Base for Tobacco Harm Reduction.* Washington, DC: The National Academies Press; 2001.
37. Institute of Medicine. *Ending the Tobacco Problem: A Blueprint for the Nation.* Washington, DC: The National Academies Press; 2007.
38. Gartner CE, Hall WD, Vos T, Bertram MY, Wallace AL, Lim SS. Assessment of Swedish snus for tobacco harm reduction: an epidemiological modeling study. *Lancet.* 2007;369:2010–2014.
39. Gottlieb S, Zeller M. A nicotine-focused framework for public health. *N Engl J Med.* 2017;377(12):1111–1114.
40. Ramström L. The case of snus. *Presentation at: 2nd Scientific Summit on Tobacco Harm Reduction: Novel products, Research and Policy [No Smoke Summit 2019]; May 29, 2019. Athens, Greece.* https://doi.org/10.13140/RG.2.2.25590.04164.
41. Ramström L. Sweden's pathway to Europe's lowest level of tobacco-related mortality. *Tob Induc Dis.* 2018;16(suppl 1):A607. https://www.researchgate.net/publication/323493113_Sweden's_pathway_to_Europe's_lowest_level_of_tobacco-related_mortality.
42. Warner KE, Slade J, Sweanor DT. The emerging market for long-term nicotine maintenance. *JAMA.* 1997;278(13):1087–1092.
43. Warner KE, Peck CC, Woosley RL, Henningfield JE, Slade J. Treatment of tobacco dependence: innovative regulatory approaches to reduce death and disease: preface. *Food Drug Law J.* 1998;53(suppl):1–8.
44. Warner KE. An endgame for tobacco? *Tob Control.* 2013;22:i3–i5.

Preface

When Brian Thomas, the series editor of Emerging Issues in Analytic Chemistry, asked me to submit a chapter for this volume, I surprised him—and even myself—by agreeing not only to participate as an author but to be the guest editor. It has been a journey! My first and only personal use of a smokeless tobacco (ST) product was at the National Institute on Drug Abuse Intramural Research Program. I was working in Jack Henningfield's Clinical Pharmacology Laboratory in a project to understand the effects of ST product pH on nicotine absorption and other effects. My collaborators on the study were Tom Cargulio, a pharmacy student intern, and my longtime friend and collaborator Reggie Fant. We were trying to perfect a dosing scheme for moist snuff and were experimenting with using concentric blood collection test tubes to create a plug suitable for a unit dose of about 2 grams for our research subjects. In the spirit of experimentation and in a moment of bravado, I placed a plug "between cheek and gum," just as instructed by so many popular media ads. Within a minute, I was running for the restroom—feeling the head rush, intense salivation, a wave of tachycardia, perspiration, and fighting back emesis. It was so cholinergic!! A memorable introduction to ST, a product that has been a part of the tobacco market for centuries, has adversely affected the health of millions of people, and has been a subject of my experimental and clinical research for the past two decades.

For many years, the discussion of tobacco use was centered on cigarettes, often to the neglect of other combustible and noncombustible means of nicotine delivery. However, tobacco use preceded the popularization of cigarette smoking, which began with the development of the cigarette rolling machine in 1884. In 1907, the US cigarette production (5 billion sold) was less than cigar production (7 billion sold), but by 1928 cigarette production reached 100 billion. Before its "discovery" by Columbus in the New World on or about October 15, 1492, tobacco was cultivated and used by indigenous people in both combustible and noncombustible form. There are records of tobacco powder being insufflated through tubes into the nose, dipping of snuff, and smoking of cigars. Cigarettes peaked in the United States in 1964, when over half of the adult population smoked them. The Luther Terry report in 1964 declared that cigarette smoking was a health hazard and its use was associated with lung cancer. In the decades that followed, intense public health efforts, epidemiologic research, and the realization that smoking harms the user and those in his vicinity led to laws, policies, and restrictions that eventually decreased prevalence to its current level of less than 20% of US adults.

As cigarettes sales have decreased, the use of other nicotine delivery methods has increased, particularly cigars, cigarillos, and little cigars among some demographic groups. The introduction of electronic cigarettes (e-cigarettes) and the huge increase in their use has been documented over the past decade. Other tobacco products such as lozenges and oral nicotine delivery products such as the Verve series (discontinued in early 2019) have been introduced. Pharmaceutical products that deliver nicotine in the form of gum, transdermal patches, and nasal spray were developed to facilitate smoking cessation.

This volume provides an overview of research on ST products, their place in the history of tobacco use, their effects, and their role in the current environment of harm reduction. Broadly defined, any noncombustible tobacco product or nicotine delivery system might be considered smokeless tobacco. The usual forms are chewing tobacco, moist snuff, Swedish style snus, and dry snuff. The leading product in the United States is moist snuff. In Sweden, a particular form of processed snus is ordinarily used, whereas in Africa and Southeast Asia the products are frequently made at the site of sale and contain compounds such as areca nut (in betel quid), lime, and a leaf wrapping. Products of various categories are used by millions of people worldwide. Use is especially evident among women in developing countries. ST is known to cause dental, cardiovascular, and reproductive pathology, and oral and pancreatic cancer. The public health burden is considerable, and the economic consequences are especially pernicious in developing countries, where regulatory authorities are least likely to respond.

The chapters of this volume take various points of view. The diversity of product design and marketing has implications for regulation. Epidemiology domestically and worldwide illustrates the magnitude of the problem. Laboratory analyses emphasize tobacco-specific nitrosamines and a host of other constituents and show how state-of-the-art knowledge can be used to set the standards for regulation. Laboratory results reveal the mechanisms of exposure and the rapid, robust delivery of the addictive compound nicotine. Treatment strategies to help those wishing to quit ST use are reviewed, as are product modifications for harm reduction. A discussion of the policy and regulation of ST products addresses some of the outstanding and current research questions. For example, as this book was going to print, the US

Food and Drug Administration approved a Swedish-style moist snuff (General brand snus) to be sold in the United States with a modified risk claim. This is the first time a tobacco product has been given approval to advertise and be promoted as having lower health risk. The comprehensive approach adopted for this volume provides perspective to students, clinicians, researchers, and policy personnel who confront the challenges of this unique group of products.

My entry into this research began as a simple clinical experiment to determine whether ST products of differing pH levels had different characteristics of nicotine absorption, cardiovascular effects, and subjective effects. Since then, the field has been expanded and refined by many of the contributors to this volume. We hope that our efforts inspire further curiosity, research, discussion, and science-based policy and regulation on this fascinating group of products.

Wallace B. Pickworth
Baltimore, MD, United States
October 24, 2019

CHAPTER ONE

Introduction

Wallace B. Pickworth
Battelle Memorial Institute, Baltimore, MD, United States

This volume provides an up-to-date review of smokeless tobacco products in the rapidly changing consumer tobacco marketplace. In the past 20 years, there have been dramatic changes in the policies and regulation of tobacco. Most of it has been directed at cigarettes, but there is a continuing interest in other products, including smokeless tobacco (ST). The Framework Convention on Tobacco (FCT), a World Health Organization (WHO) initiative, and the Family Smoking Prevention and Tobacco Control Act (FSPTCA) are among the international and domestic laws that have begun science-based regulation of the tobacco market. In the United States, the FSPTCA established a new agency, the Center for Tobacco Products (CTP), within the Food and Drug Administration (FDA). The CTP has broad and strong regulatory authority over all tobacco products. Their authority over ST was defined in the original legislation in 2009 and was extended to other tobacco products, including electronic cigarettes, in 2014. A stated goal of the legislation is to use science-based results to inform regulations that serve the public health. The material in this volume contributes to the discussion.

Two other factors have brought change to the tobacco marketplace. First, there has been an increasingly visible adaptation of harm reduction approaches to the regulation of tobacco products. Harm reduction implies the acceptance of some health risk to reduce overall risk. All tobacco products are associated with health risks, but there is considerable evidence that the risk is not equal. Combustible tobacco poses the most risk, and the cigarette is the most pernicious form. At the lower end of the continuum are the slow-release nicotine-only pharmaceutical delivery products such as the transdermal patch and nicotine polacrilex gum. ST products carry acknowledged risks of addiction, mouth, head, and neck cancer, and cardiovascular and reproductive effects; but relative to cigarettes, they have a much lower overall risk profile. A harm reduction approach would suggest that ST products should be less regulated and less taxed and have fewer and less restrictive control regulations than more harmful products.

Smokeless Tobacco Products
ISBN: 978-0-12-818158-4
https://doi.org/10.1016/B978-0-12-818158-4.00001-7

The tobacco marketplace has been rocked by the recent introduction of electronic cigarettes or electronic nicotine delivery systems (ENDS) and heat-not-burn products. ENDS affect nicotine delivery by heating solutions of nicotine in glycerol or polyethylene glycol, whereas heat-not-burn products rely on the low-temperature warming of tobacco. The emission from ENDS or heat-not-burn products is substantially lower in harmful compounds than tobacco smoke of conventional cigarettes. It would seem that these products might lower the health risk to the user and diminish exposure to others by reducing secondhand smoke. However, electronic cigarettes are relatively new, and their health risks may not be fully evident. Furthermore, they have already gone through three or more generations of products, so the health hazards of each variant are subject to individual study. Heat-not-burn products have been on the international market about 5 years, though only recently approved for sale in the United States. There are very little data on health consequences of their long-term use. In contrast to ST, which is a market staple, new products that deliver only nicotine or nicotine with fewer other chemicals will continue to be introduced. Their long-term effect on the population use of cigarettes and other forms of tobacco is uncertain. ST, given its worldwide use that has persisted for centuries, will likely remain a significant part of the tobacco market, exerting personal and public health consequences for years to come.

This volume concentrates on many of the most actively researched areas in the field of tobacco science—particularly as they apply to ST. As summarized below, the beginning chapters review the various ST products, their epidemiology, and pharmacology. One chapter discusses the diagnosis and treatment of ST addiction. Other chapters review the chemistry in product analysis and the biomarkers of human consumption. A chapter is devoted solely to an explanation of the carcinogenic mechanisms from components of ST. The final chapter discusses regulatory policy.

Vaughan Rees and his colleagues (Chapter 2) cover the various types of ST products and their design and marketing. ST is not a singular product; there are many. This diversity must be appreciated to understand that it is difficult to discuss ST as a single nicotine delivery product. Each product has unique design features. An important implication is that there are specific attributes that can be regulated, such as flavor, packaging, pH, nicotine level, and toxicant content. Furthermore, tobacco manufacturers, in an attempt to retain existing customers and gain new ones, have introduced innovative products and marketing techniques. Cigarette manufacturers have now entered the ST market with similar objectives and strategies. This chapter addresses those innovations and how they must be understood in designing regulatory policy to discourage initiation of tobacco use in any form or, for persons who will not quit, to encourage a switch to putatively less-toxic forms. Types of products and the regional and demographic characteristics of use in the various markets are given. Innovations are evaluated for their effectiveness. These factors are discussed in terms of how they inform policy and regulatory strategies for an evolving tobacco market.

Gretchen McHenry and her team provide a comprehensive review (Chapter 3) of the epidemiology of ST use in the United States and internationally. Global use is estimated to be nearly 300 million people, much of it in lower income, developing countries and among women. In the United States, about 2% of adults are users, but they are not uniformly distributed in the population; incidence is highest in men, rural states, Native Americans, and youth. That review reminds us of the vast number of women who use the products in Africa and Asia, where smoking among women is culturally disapproved. Unfortunately, use in women has been associated with poor birth outcomes and other recognized health risks. Epidemiologic data are important indicators of use trends, the effectiveness of regulation, and the influence of new products on the market. Some of the risks are well known and acknowledged by users, including addiction, dental problems, and oral cancers, but other established risks such as cardiovascular and circulatory disease, reproductive effects, and pancreatic cancer are underappreciated or even unrecognized. Underappreciation may account for persistent use. Data such as these could inform health messaging that promotes prevention of ST initiation and advocates cessation.

Findings from laboratory studies are presented in Chapter 4. The relatively few such studies have investigated the role of the product pH in nicotine absorption. Other studies have investigated flavor as a determinant of use and appeal. The format of moist snuff, as loose tobacco or in small "teabag" sachets, has been investigated to determine whether such design changes alter rate of nicotine absorption. A central tenet of the research is that the speed of nicotine delivery is related to abuse liability (addictiveness). The chapter reviews self-report questionnaires that assess use patterns and dependence potential. Data from subjective questionnaires can be used by clinicians to guide and assess therapeutic decisions and by researchers to assess the effects of product manipulation. Findings from laboratory studies provide empirical evidence that can be used for science-based regulation.

Herbert Severson and his colleagues (Chapter 5) review the literature on the treatment of those wishing to quit ST. The chapter covers the measures of nicotine dependence that are specifically developed for ST users. Drugs, behavioral assistance, and counseling have been employed. ST users as a group are surprisingly treatment-resistant, in many cases more difficult to treat than cigarette smokers. Unique interventions specific to ST use are being developed and tested. In general, research has yielded good evidence and support for several methods, with some outcomes exceeding those for smoking cessation. However, refinement of existing methods and delivery systems is needed, as well as innovation in methods. Future directions in which studies and therapeutic applications might go are projected.

Irina Stepanov and Dorothy Hatsukami (Chapter 6) provide a comprehensive review of the chemical characterization of ST products and biomarkers of use in humans. They review the absorption and metabolism of nicotine and the toxicant exposure to tobacco-specific nitrosamines (TSNA). The chapter is an overview of sources, levels, and the variation of harmful constituents. Established biomarkers that assess exposure to ST constituents are discussed, and the analytical methodologies used for the measurement of constituents and biomarkers are reviewed. Variations in the levels of key harmful constituents across products and the impact of these variations on related exposures are emphasized, and policy implications are discussed. The products are characterized by enormous diversity of types and formulations, from uncured dry tobacco that is used by itself to complex recipes made with various additional ingredients that can modify addictiveness, toxicity, and carcinogenicity. Adding to this complexity is the diversity of tobacco plant types and processing. Data on exposure to chemical toxicants and carcinogens from the use of ST products is scarce and mostly limited to the United States. However, cancer risks worldwide mirror the variations in the levels of key known carcinogens in products. In India, many products contain high levels of certain carcinogens, and incidence in head and neck cancers is remarkably high, while Swedish snus contains low levels and the associated cancer risk is virtually nonexistent. Therefore, chemical characterization of products and related exposure in users is key to the development of preventive measures and science-based product regulation.

In Chapter 7, Steven Hecht explores the carcinogenic mechanisms of the TSNA found in ST. They are the most abundant and potent carcinogens in ST and are acknowledged by the International Association on Cancer Research (IARC) as Class 1 carcinogens in humans. They appear especially responsible for tumors of the mouth, esophagus, and pancreas. Interestingly, they are not biologically active as they exist in the product, but must be metabolized by cytochrome enzymes in the liver and elsewhere to produce the toxic metabolites. One of the enzymes that catalyze that conversion is CYP 450 2A6, which is responsible for the metabolism of nicotine to cotinine and 3-OH cotinine. Ironically, the enzymes involved in the metabolism and deactivation of the addictive component of tobacco, nicotine are responsible for the activation of other components that lead to tumors. The recognition of the carcinogenic potency of the TSNA has led to recent calls to limit their concentration in ST to less than 1 ng/g dry weight.

The question of how growing, manufacturing, processing, and packaging influence levels of TSNA is reviewed by Stephen Stanfill in Chapter 8. This chapter covers the sources of toxicant compounds and their precursors. There is an extensive discussion of the measurement of heavy metals, TSNA, and additives that contribute to the overall toxic profile. Tobacco absorption of soil nitrate, microbial nitrate reduction, and nitrosation are thought to contribute to the formation of TSNA, including potent carcinogens. The encouraging message from this review is that many of the toxicants in ST can be limited or eliminated by achievable agricultural or practical changes to manufacturing process.

The volume concludes with a discussion (Chapter 9) on regulatory policy. Marc Parascandola and I review the provisions of the FCT and FSPTCA specific for ST regulation. Case studies illustrate the counter-advertising and heath messaging that attempt to diminish ST use. Other health promotion efforts that have been implemented are considered. As with other tobacco products, discussion of ST products is complex, because they are at once drugs, legal commercial goods, income sources for their producers and the entities that tax them, and cause of significant adverse health outcomes and public health burden. Sadly, much of the personal and public health consequences are borne by people in the most impoverished countries. The long history of ST use and the current enthusiastic commercial marketing suggest that these harmful products may be in the discussion of tobacco and health for years to come. It is our hope that the contents of this volume will inform the on-going discussion of ST and inspire further research that directly addresses questions of importance to understanding and regulation of these products.

CHAPTER TWO

Smokeless tobacco product design and marketing: targeting new populations in a changing regulatory environment

Vaughan W. Rees[1], Olalekan A. Ayo-Yusuf[2], Richard J. O'Connor[3]
[1]Harvard T.H. Chan School of Public Health, Boston, MA, United States
[2]Sefako Makgatho Health Sciences University Medunsa, Pretoria, South Africa
[3]Roswell Park Comprehensive Cancer Center, Buffalo, NY, United States

Introduction

Smokeless tobacco (ST) has a long history and is the predominant form of tobacco in some global regions, yet it has garnered a somewhat lower priority for research and policy development than combusted forms. Some ST products have lower health risks than smoking because they do not expose consumers to many of the highly toxic chemical by-products of combustion. Still, ST use causes serious health problems, including increased risk of cancers of the head and neck.[1] The global ST market has undergone important changes in the past two decades, as combusted tobacco has become more heavily regulated in developed countries and consumers seek lower-risk alternatives. In response, cigarette manufacturers have entered the ST market with novel ST products that offer putative lower risk and greater personal convenience. The capacity of the tobacco industry to innovate has been well documented,[2,3] yet relatively less attention has been given to recent innovations in ST. This chapter will consider new developments in product design and marketing used by manufacturers to gain new consumers and retain existing ones, often through targeting subpopulations. By focusing on factors used by manufacturers to increase appeal rather than on health risk factors, this chapter will consider how tobacco control regulatory initiatives must address industry strategies. The goal of regulation should be not just to protect the health of current consumers but to prevent people from ever initiating use. In developed countries with a tradition of ST use, such as the United States and Scandinavian

Smokeless Tobacco Products
ISBN: 978-0-12-818158-4
https://doi.org/10.1016/B978-0-12-818158-4.00002-9

countries, demand for combusted tobacco is decreasing in favor of noncombusted products. Thus, a deeper understanding of strategies used by ST manufacturers to design and promote their products can inform future tobacco control regulations intended to reduce consumer demand for ST, while identifying opportunities to lower the health risks for adult smokers with properly regulated, reduced-risk ST products.

Types of products

ST is available globally in myriad forms that are used orally or nasally. Table 2.1 gives information on common forms, and Fig. 2.1 shows some examples. As with any product, ST is manufactured to meet the preferences and expectations of target consumer groups. Some forms are traditional, made with limited technology under rudimentary conditions. Others are technologically innovative, designed to attract consumers in competitive developed markets. Traditional forms occur mainly in developing countries in the Indian subcontinent, parts of Africa, and the Middle East. In India, which historically has had the largest ST industry by volume,[11] use exceeds that of combusted tobacco.[12] Popular products are dry snuff (e.g., bajjar, also known as tapkir), a finely ground tobacco powder used orally or nasally, and gutkha, khaini, and naswar ("nass"), which are made of dried and chopped tobacco blended with other ingredients including crushed areca nut, slaked lime, ash, and flavors such as catechu extract, cardamom, and menthol. Also popular is a powdered tobacco paste that is applied to the teeth and gums, sometimes as a dentifrice; common forms are gudakhu, gul, and mishri (Fig. 2.1). Mawa, zarda, kiwam/quiwam, and shamah blend flavorings and binders such as ground areca nut, lime, oils, and spices; this form is chewed or placed between the cheek and gum to deliver nicotine via the oral mucosa. In North Africa, dry snuff formulations such as naffa (also called tenfeha or nufha) are placed inside the lip. In southern Africa, a popular form of traditional snuff is a mix of powdered tobacco with charred plant and/or ash, which is alkaline and acts as a buffer to facilitating nicotine absorption. In the Sudan region, toombak is a moist product made from sundried tobacco that is fermented, mixed with sodium bicarbonate, and cast into small balls for oral use. Other countries have specific ST variants, such as maras in Turkey, which is powdered sundried tobacco mixed with oak or grape leaves, and chimó in Venezuela, a hardened paste made from crushed, boiled tobacco leaf mixed with sodium bicarbonate, sugar, ashes, and flavorings (Fig. 2.2).[5,13]

Table 2.1 Major types and characteristics of smokeless tobacco.

			Product characteristics				
Product type	Brand names	Manufacturers	Total nicotine concentration (mg/g)[5–8]	Free nicotine concentration (mg/g)[5–7]	Common flavors/ Flavorants[4–10]	Preparation and formulations[5,6,4,9,7,8,10]	WHO region[1,4]
Chewing tobacco	Red Man, Days Work, Apple, Brown, Natural Leaf, Union Standard, Tinsley, WNT, Levi Garrett, Taylors Pride, Cannon Ball, Moore's Red Leaf, Cumberland, Mammoth Cave, Cotton Boll, Kentucky, Warren County, Rough Country	Swedish Match North America, American Snuff Company	Plug variety: 5.1−15.1 Twist variety: 21.6−40.1 mg/g	Plug variety: 0.01−0.04 Twist variety: 0.02−0.22 mg/g	Licorice, sugar	Cured tobacco treated with leaf extract, flavored, and dried	Americas (primarily United States)
Moist snuff (dip)	Copenhagen, Skoal, Red Seal, Husky, Grizzly, Kodiak, Kayak, Redwood, Gold River, Silver Creek, Cooper, Silverado, Tim bar Wolf, Longhorn, Red Man	Conwood Company, National Tobacco Company, Swisher International, Swedish Match North America, US Tobacco	4.42−25.0 Mean of US brands: 12.3	Mean of US brands: 4.0	Mint, wintergreen, fruit, cinnamon	Cured, fermented, and flavored tobacco, fine or long cut; pouches or loose	Americas, Europe
Snus	General, Catch, Ettan, Grovsnus, Göteborgs Rapé, Kronan (Swedish Match); Lucky Strike, Pall Mall, du Maurier (British American Tobacco); Camel (R.J. Reynolds); Marlboro (Philip Morris); Skoal (U.S. Smokeless Tobacco Company); Knox, Skruf (Imperial Tobacco), Tobaccorette	R.J. Reynolds, Philip Morris, Swedish Match North America, US Tobacco	Sweden: 12.8−28.2 Mean of US brands: 10.46	Mean of US brands: 3.08	Mint, wintergreen, fruit, cinnamon, molasses	Pasteurized finely cut tobacco; pouches/sachets or loose	Americas, Europe (especially Scandinavia)
Dry snuff	Levi Garrett and Sons, Dental, Honest, Peach Sweet, Tube Rose, W.E. Garrett & Sons, Silver Dollar	American Snuff Company, Kretek International, Inc.	In US: 4.7−24.84	In US: 0.03−3.13		Fire-cured and fermented tobacco with added flavors; powder	Americas, Africa, Europe, South-East Asia
Bajjar/ tapkir	Typical cottage product or home-prepared		na	na	Menthol, floral	Roasted and powdered tobacco used as dentifrice	Americas, Europe, South-East Asia (primarily India)

(Continued)

Table 2.1 Major types and characteristics of smokeless tobacco.—cont'd

			Product characteristics				
Product type	Brand names	Manufacturers	Total nicotine concentration (mg/g)[5–8]	Free nicotine concentration (mg/g)[5–7]	Common flavors/ Flavorants[4–10]	Preparation and formulations[5,6,4,9,7,8,10]	WHO region[1,4]
Naffa/ tenfeha/ nufha	Typically cottage product or home-prepared		na	na		Dry snuff; used nasally	Africa
Gutka	Manikchand, Moolchand, Tulsi, Shimla, Parag, Sir, Goa, and Sikandar		0.16–4.20		Sweeteners or savory flavoring agents	Areca nut, slaked lime, and powdered tobacco	South-East Asia
Khaini	Raja Kuber, Wiz, Buddha Lal, Chaini, Raja Chap, Ansul Tobacco, Mirage, Ganesh tobacco 701, Patta Chhap Tei Tobacco		2.53–4.79	2.48–4.68		Areca nut, slaked lime, and sun-dried or fermented tobacco leaves; paste	South-East Asia
Naswar	Three Star, Wail Zaman, Sardar and Irfan, Lachiwaja	Karachi, Bannu, Swabj, Mardan, Charsadda, Quetta, Jhob, Mohamand	11.8–28.7	8.84–13.2	Cardamom, menthol	Dried tobacco, ash, colorants, oils; balled for oral use	Africa, Eastern Mediterranean, Europe
Gudakhu	Natraj		na	na	Molasses	Tobacco leaf dust, molasses, red soil; paste used as dentifrice	South-East Asia
Gul	Shajadi Gul, Mujamal Hussain Musarrf Bahi Shahi Eagle, Md. Mustafa Asgar AliGul		33.4–34.1			Pyrolized tobacco with tendu leaves; used as dentifrice	South-East Asia
Mishri	Typically cottage product or home-prepared		2.73	0.09		Baked or roasted tobacco; powder used as dentifrice	South-East Asia
Mawa	Typically cottage product or home-prepared		0.16–4.20	0.11		Areca nut, sun-cured tobacco flakes, and slaked lime	South-East Asia

Zarda	Baba, Baghban Zafrani Zarda, Ratna Zafrani Patti, Gopal (India); Zahoor Zafrani Patti, Raja Jani Zafrani Patti, Sunbrand Zafrani Banarasi Patti, Shahzadi Zafrani Patti, Najma Zaffran Patti (Pakistan); Dulal Mishti, Hakim Puri, Bat One Baba, Bullet, Surma (Bangaldesh)		14.6−65.0 Mean in India: 30.43	Mean in India: 0.05	Saffron, menthol	Chewing tobacco flakes with spices, dyes, lime, sometimes areca nut; chewed	South-East Asia, Eastern Mediterranean
Kiwam/ quiwam	Avon, Kashmiri, Nauratan, Raj Ratan, Pradip		na	na	Cardamom, saffron, aniseed, musk	Boiled tobacco leaves with flavors and additives; paste	South-East Asia
Toombak	Typically cottage product or home-prepared		9.56−28.2	5.16−10.6		Tobacco mixed with baking soda and water; balled for oral use	Africa (primarily Sudan, Chad)
Shamah/ shammah	Typically cottage product or home-prepared		na	na		Powdered tobacco, slaked lime, ash, black pepper, oils, and bombosa; powder or paste often wrapped in paper	Eastern Mediterranean, Europe
Maras	Typically cottage product or home-prepared		na	na	Oak, walnut	Powder of wood ash, dried, leaves, tobacco, and water	Europe (primarily Turkey)
Chimó	El Tovareño, El Tigrito, El Sabroso, El Gran Búfalo, El Dragon, El Morichal, San Carleño		5.29−30.1	1.32−27.4	Cocoa, brown sugar, vanilla	Tobacco leaf combined with baking soda, Mamón tree ashes, and flavorings; hardened paste	Americas (primarily Venezuela)
Creamy snuff	IPCO, Dentobac, Tona, Ganesh, Charotar, Musa Ka, Rehmat Khan, Chad Tara, Dulhan, Suraj, Asif Ka	Asha Industries, Goran Pharma Ltd.	5.62−10.0	2.36−3.82	Spearmint, menthol, camphor	Tobacco mixed with glycerin, aromatic substances, water, and oil; used as dentifrice	South-East Asia

Note: List of products is not exhaustive.

Figure 2.1 Some examples of smokeless tobacco products used in India: khaini (upper left), gutkha (lower left, left side), mishri (lower left, right side), and gul (right). *Photos courtesy of Clifford Watson, Centers for Disease Control and Prevention.*

In developed countries, commercially manufactured products dominate consumer markets and are concentrated in the United States and Scandinavia. For the first half of the 20th century, chewing tobacco was the most popular ST in the United States. It comes in loose leaf, pellet, and plug forms and is comprised of tobacco blended with sweetener and flavorings. In

Figure 2.2 Some examples of smokeless tobacco products used in other global regions: mawa (upper left), zarda (lower left, left side), chimo (lower left, right side), and quiwam (right). *Photos courtesy of Clifford Watson, Centers for Disease Control and Prevention.*

recent decades, moist snuff, also called dip or chew, became dominated in the United States.[14] The tobacco is air or fire cured and aged before being cut, blended, fermented, and flavored. The form is based on the texture of the tobacco cut (long or fine) and flavor. The variant called snus has a lower moisture content and is made from finely cut tobacco that is pasteurized instead of fermented. Snus products, which originated in Sweden, are often portioned into small cellulose fiber pouches, each pouch a measured portion that makes placement between lip and gum convenient. In the past decade, dissolvable forms of oral products made from compressed, powered tobacco have been developed to appeal to health-conscious consumers. While earlier iterations of dissolvable tobacco lozenges did not meet with market success, new forms have recently been introduced in the United States. The range of ST designs used by manufacturers to enhance consumer appeal are discussed later in this chapter.

The evolving smokeless tobacco market

In 2017, the value of the global ST market was estimated at US$12.85 billion. Over 131,000 tonnes of tobacco were processed in some 18 countries.[11] Yet despite its size, ST represents just 1.6% of the total global tobacco market, and its value was surpassed in 2017 by that of the burgeoning e-cigarette (vaping) market. However, ST still dominates in some regions, such as India, where traditional products maintain a higher prevalence than combusted products.[15] The World Health Organization (WHO) South-East Asia region (Bangladesh, Bhutan, Democratic People's Republic of Korea, India, Indonesia, Maldives, Myanmar, Nepal, Sri Lanka, Thailand, and Timor-Leste), notably the Indian subcontinent, contains 90% of the world's 250 million ST consumers.[16] Moreover, the market has grown faster than that for cigarettes in some areas: in the United States, by 6.9% annually from 2003 to 2017, compared to 0.4% for cigarettes.[11] The US ST market was estimated at US$7.2 billion in 2018, making it the world's largest in value.[11]

The growth of the US market may be attributable to several factors, which are underpinned by an increasingly regulated cigarette market and changing consumer preferences. Tobacco control initiatives have been adopted in recent decades to communicate the risks of combusted tobacco, and increased knowledge of these risks, combined with increasing restrictions on indoor smoking, has contributed to making smoking less socially acceptable. Yet the fast-changing landscape of global tobacco control has not been applied equally to all types of products. For example, although a majority of countries have implemented specific actions required by the WHO's Framework Convention of Tobacco Control (FCTC) relating to combusted products, most countries have not actively advanced efforts to curb ST use.[17] This inequity in implementation of prevention policies may have unintentionally led to greater opportunities for ST manufacturers to market their products to new consumers, including youth and health-conscious consumers looking for alternatives to combusted tobacco.[18]

Innovations in smokeless tobacco products: design and marketing strategies

Throughout the last century, tobacco manufacturers have sought to gain competitive advantage by the use of innovation in the way their products are designed and marketed. Product innovation typically involves sophisticated, highly researched changes in design and performance to attract new consumers and maintain existing ones. In turn, innovations in ST follow this pattern and have generally been aimed at enhancing two things: abuse liability and product appeal. The potential to produce dependence, or abuse liability, is the likelihood that a given product will result in a consumer becoming addicted and is achieved primarily through the mechanism of nicotine delivery.[19,20] Product appeal is a function of nonnicotine characteristics that enhance attractiveness, social acceptability, and ease of use. Traditionally, ST manufacturers were not as closely associated with product innovation as cigarette manufacturers, who relied heavily on it in the 20th century to give rise to the modern cigarette.[21] ST manufacturers took a more limited approach, such as the use of flavorings and variation in nicotine delivery.[22] However, since around the year 2000, novel ST products that reflect a substantial shift in the approaches used to recruit and retain users have been introduced.

Developments since the mid-2000s have transformed the ST market in the United States, as leading cigarette companies acquired ST companies and introduced new ST products. Strategically, these moves have aimed to attract current smokers and never smokers with ST products that purport to present lower health risks while ensuring ease and convenience of use. R.J. Reynolds, a leading cigarette manufacturer, acquired Conwood, the second largest ST manufacturer in the United States, in 2006, and Altria acquired U.S. Smokeless Tobacco Company in 2009. These acquisitions placed control of leading ST moist snuff brands Kodiak and Grizzly (Conwood) and Skoal and Copenhagen (U.S. Smokeless Tobacco Company) in the hands of cigarette manufacturers, ostensibly to synergize marketing of cigarettes and ST products. These companies also introduced potential reduced risk snus products marketed under leading cigarette brand names, including Camel Snus and Marlboro Snus, and dissolvable products sold as Camel Dissolvables (Orbs, Sticks, and Strips) (Fig. 2.3). Data from Massachusetts show that the number of snus brands sold in that state increased from 4 in 2003 to 62 in 2011, before dropping to 26 in 2012, indicating a high point in snus product marketing.[6] Very recently, Altria acquired an 80% stake in Swiss ST manufacturer Burger Sohne, that produces a snus product marketed as On![23] These actions reflect a general trend toward expansion of the noncombusted market, as evidenced by Altria's acquisition of a 35% stake in the manufacturer of the JUUL e-cigarette and the pending introduction of Altria's heat-not-burn product, IQOS.[24,25]

Innovations in design are promoted in sophisticated ways to optimize consumer appeal and may subvert existing tobacco control regulations.[26–29] Therefore, control authorities must surveil the entire ST market to understand physical design, formulation, and constituents of products and the methods used by manufacturers to communicate to both old and new consumers.[30] Insights on innovations will inform potential regulatory strategies to prevent the emergence of new fronts in the ongoing tobacco epidemic by preventing initiation of ST use while providing lower risk alternative for adult smokers.

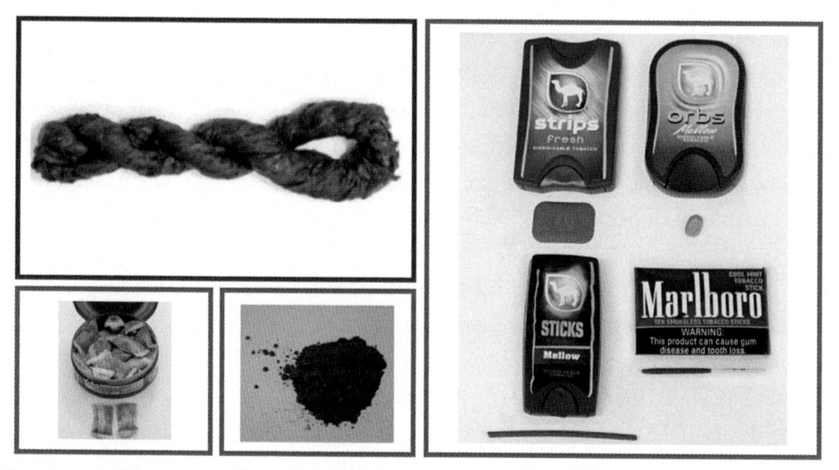

Figure 2.3 Some examples of smokeless tobacco products used in the United States: chewing tobacco twist (upper left), snus pouches (lower left, left side), dry snuff (lower left, right side), and dissolvable oral tobacco (right). *Photos courtesy of Clifford Watson, Centers for Disease Control and Prevention.*

Design features that promote use

For decades, tobacco manufacturers have systematically manipulated and continue to refine the design and formulation of their products to enhance consumer appeal.[31–33] ST manufacturers do likewise to meet the preferences of current consumers and to increase interest among targeted groups of potential consumers.[5] These features ultimately influence consumers' preferences by enhancing product appeal and potentially promoting tobacco dependence.[34]

ST manufacturers seek to enhance dependence—that is, abuse liability—by optimizing the speed and amount of nicotine dosing while providing appealing chemosensory characteristics such as taste, coolness, and smoothness to make the product easier and more pleasant to consume.[22,35–38] Three major areas of innovation—manipulation of nicotine delivery, product formulation, and flavoring—are considered in the following sections.

Nicotine

Nicotine is a leading feature of tobacco product design and provides the primary reinforcing effect that underpins dependence. Its effects are perceived by consumers as pleasurable (after sometimes negative symptoms from early exposure), which directly influences preferences and use behavior.[36] While ST delivers nicotine more slowly than smoking, as measured via blood plasma level,[37] overall delivery is about the same and is sufficient to promote and maintain dependence.[37]

Nicotine level varies substantially between and within products. The form of nicotine also can be varied by adjusting the proportion of that exists as free-base (unprotonated), which more readily permeates mucous membranes, including the buccal mucosa and the blood–brain barrier. This causes a more rapid onset of effect, thus supporting the development and maintenance of dependence.[38] The proportion of free nicotine is a function of pH. Higher (more alkaline) pH yields a greater proportion of free-base.[39,40] pH is readily altered by the addition of chemical buffering agents and salts, and there is extensive evidence that manufacturers have used this approach to modify product abuse liability.[41]

Since at least the 1980s, manipulation of free nicotine has been a mainstay of design.[38,42,43] Manufacturers made products, notably moist snuff, to provide variations in nicotine dosing that were convenient to consumers. This was done to advance the so-called graduation strategy, whereby low-nicotine moist snuff was used to attract novice users who had not developed tolerance to high nicotine doses. The rationale was that nicotine is aversive in novices, and lower doses are more effective at promoting initiation because the product can be more easily and comfortably consumed. Some brands, such as Skoal Bandits, had nicotine levels as low as 7.5 mg/g, compared with 10.3—11.4 mg/g for other brands.[43] Products with iteratively higher doses are available to more experienced users to support increasing tolerance and dependence.[44] Nicotine levels in moist snuff sold in the United States now range from about 6 to 23 mg/g,[6,45] which is wide enough to suggest that there is considerable variation in consumer preference for nicotine levels.

Manufacturers have also manipulated physical characteristics to appeal to smokers, apparently to promote dual use with cigarettes rather than to support complete switching. Seidenberg and colleagues found that Swedish style pouched snus sold in the United States had a far lower concentration and proportion of unionized nicotine than that sold in Sweden.[46] While lower levels of free nicotine would be less likely to promote dependence and thus support higher use, the authors speculated that a lower abuse liability might be designed by manufacturers of American snus to promote dual use with smoking. Cullen and colleagues found that free nicotine in snus products sold in Massachusetts increased between 2003 and 2012 at an overall rate of 0.19 mg/g per year, with this increase driven by products made by Swedish Match North America (General Snus) and Reynolds (Camel Snus).[6] During the same period, free nicotine levels in moist snuff were relatively stable, although increases were seen in mint- and menthol-flavored and pouch products, as well as those made by U.S. Smokeless Tobacco Company (e.g., Copenhagen, Skoal).

Formulation

The range of formulations has evolved, giving rise to a variety of products targeted at subgroups of consumers. Companies in Sweden and the United States have varied the manner in which tobacco is prepared, packaged, and presented. For example, it may be dried and cured, which achieves a lighter, smoother quality, or fermented, which increases sweetness. Most moist snuffs in the United States are fermented and chopped or shredded in different grades to achieve a "cut" of a finer or longer form. Long cut tobacco is easier to manipulate into a wad for placement inside the cheek and has longer-lasting effects. Finer cuts are less convenient but have more rapid effects, owing to a greater tobacco surface area in contact with mucosa. Manufacturers have also used design innovation to address health concerns among smokers and have developed products to reduce the perception of risk relative to smoking.[47] For example, Swedish snus is pasteurized by steam, resulting in lower concentrations of carcinogenic, tobacco-specific nitrosamines.[48,49] The GothiaTek standard used by Swedish Match requires that concentrations of *N*-nitrosonornicotine (NNN), 4-(methylnitrosamino)-1-(3-pyridyl)-1-butanone (NNK), lead, and aflatoxins not exceed specified limits, although these industry-established limits have been criticized as insufficient to protect the health of consumers.[50] Another variant is the range of powdered, compressed tobacco dissolvable products in the form of lozenges, tablets, toothpicklike sticks, and strips that resemble popular breath fresheners.

Spitless products

One targeted form is spitless moist snuff. Chewing or sucking on older products stimulates salivary excretion, which produces a dark, pungent fluid that cannot be swallowed because of irritation to the esophagus and gastrointestinal discomfort. Spitting is aversive to many users and a source of social stigma.[51] Lowering the moisture content and adding salt reduces saliva production. Swedish-style formulations such as Camel Snus are marketed as spitless to appeal to smokers and new potential ST users.[52]

Pouched products

Pouches are used primarily for finely ground snus. The user has the convenience of a premeasured portion of tobacco corresponding to a standard nicotine dose that is easy to place between the cheek and gum. The pouch material minimizes the discomfort of fine tobacco particles becoming embedded between the teeth or migrating to other parts of the mouth. The pouch is also less visible than loose snuff, and concealment of use may reduce the social stigma. The benefit is evident in marketing to US smokers, who have historically regarded ST as socially undesirable.[53,54]

Powdered, compressed tobacco

Commonly known as dissolvables, compressed tobacco is a recent innovation, having been introduced to the United States in 2001 as the Ariva lozenge, followed in 2003 by Stonewall. They dissolve in the mouth, a convenient and more socially acceptable means of consuming tobacco. Not surprisingly, they were positioned by their manufacturer, Star Scientific, as lower risk alternatives to smoking. Reynolds American later introduced Camel Dissolvables in the variants Orbs, Sticks, and Strips, which were targeted at smokers as "adjacency" products that could be used where smoking was not possible, and suggesting lower health risk. Early concerns were expressed by the public health community about the sweet and mint flavoring and presentation of Camel Dissolvables, which resembled some candies popular with young people such as Tic-Tacs.[55] Further concern was raised about the potential for poisoning in children who accidentally consumed them,[55] although recent data suggest that this has not been substantiated perhaps because of the low prevalence of dissolvable use.[56] Likewise, the FDA's Tobacco Products Scientific Advisory Committee reported in 2012 that, while dissolvables could reduce an individual's risk of tobacco-related disease, the market penetrance was so low that estimates of population harm could not be made.[57] After a decade of poor sales, in 2012 Star Scientific withdrew Ariva and Stonewall and quit the tobacco business following an FDA ruling that disallowed claims of low health risk.[58] Camel Dissolvables were withdrawn from the US market in 2013 following low consumer acceptance and corresponding poor sales.[59] Lozenges were recently reintroduced by Reynolds American.[60]

Flavoring

Flavors make tobacco products more appealing. They are especially important for novices because plain tobacco can have harsh, smoky, or bitter qualities. Manufacturers use flavorings to minimize aversive characteristics and maximize sensory appeal. Older styles of ST were typically unflavored, although sugar and molasses were used for some chewing tobacco and moist snuff products. Wintergreen, a menthol-like cooling additive, was one of the few flavorings used in a limited number of early moist snuffs. Most current ST is flavored, and there is much variety, including fruit (peach, apple, grape), chocolate, spice (cinnamon), mint (wintergreen, spearmint, peppermint), and alcoholic drinks (bourbon, rum). Research has shown that flavor chemicals in some moist snuffs are present at higher levels than in candy.[61] A sample of 187 brands sold in Minnesota in 2013 found that 43% were flavored with mint (nonmenthol variants, including wintergreen) and 10% with fruit or sweet additives.[62] These findings were confirmed in a national sample, showing that about 60% of moist snuff brands sold in the United States are flavored.[63]

The use of flavorings has been criticized for targeting young people, including adolescents.[64] Indeed, analyses have shown that flavored moist snuff is more likely to have lower pH and free nicotine, which is a characteristic of starter products. Concern has also been raised about the use of additives that are banned from food products, such as coumarin, which has been identified in Camel Orbs.[65]

Communicating innovations: marketing

Tobacco manufacturers make extensive use of opportunities to communicate their products. Communications are intended to raise product awareness among consumers, differentiate a product from its competitors, and facilitate appeal by conveying the characteristics that may be attractive to consumers. They use several channels to achieve those goals, most notably through advertising and promotion, as well as product packaging, which reflects and reinforces broader promotional efforts at the point of use. Communications aim to create a brand image by providing subjective pictures and suggestions that build a specific profile with which individuals might personally identify. Images often used in the United States have been oriented toward masculine outdoor themes such as rodeo, fishing, and baseball.[54] Newer products such as Camel Snus emphasize more sophisticated, urbane, indoor themes aimed toward new audiences, including urban women smokers.[52,54] Recent marketing has highlighted the substitutability of ST for cigarettes, especially in the context of indoor smoking bans, and the potential for lower health risk.[64,66] Marketing is also used to communicate design features, such as nicotine level, flavoring, and presentation. Product characteristics play an important part in shaping consumers' perceptions, and internal tobacco industry research suggests that these perceptions may further influence consumers' response to the product when it is sampled.[34] This is done by packaging: Cigarette manufacturers use images, written text phrases (e.g., rich, smooth, spicy), and colors to convey information about nicotine level or flavor characteristics,[67,68] and this can be applied to ST.[69] Sales of flavored ST (primarily moist snuff and snus) increased by 72% between 2005 and 2011, underscoring the importance of these products in the ST market.[63]

Communication channels

Strategies have changed over time. In 1985, the US Federal Trade Commission reported marketing expenditure by ST manufacturers of approximately US$80 million. This sum has increased steadily, to US$718 million in 2017.[14] So too have the mechanisms for reaching targeted audiences.

Previously the appeal was to males of lower socioeconomic position, often from rural areas. In the early 1970s, radio and television were the principal media. In 1986, a federal ban on advertising of ST products by electronic media was enacted,[70] and advertising moved to magazines in conjunction with other promotional activities through the 1990s. These included sponsorship of sports events such as NASCAR motor racing and rodeos, the use of product-branded merchandise such as T-shirts, caps, and bags, advertising at the point of sale, and payments to retailers for optimal product placement in stores.[71] The Smokeless Tobacco Master Settlement Agreement of 1998 (a legal settlement between 48 US states' attorney generals and the leading ST manufacturer, U.S. Smokeless Tobacco Company) resulted in bans on outdoor billboards and restrictions on marketing involving sports, sponsorship of public events, and promotional products.[72]

Given these restrictions, strategists took advantage of the digital revolution. This took the form of company websites, targeted online advertising, and sponsored social media, which grew from US$72,000 in 2002 to US$11 million in 2017.[14] Magazines have continued to be used, especially youth-oriented publications that cover music, the entertainment industry, outdoor recreation, and sports.[73] Sweepstakes, price reductions, and coupons continue, and brand websites have been added.[74,75] Social media are leveraged, including a strong presence on YouTube.[76,77] Social media and web content are easily accessible to young people, including minors.[78] Videos that promote specific ST brands and are created by users, including those under the age of 18, have been heavily represented on YouTube.[79]

Segmentation of target markets: communicating differences

For over half a century, tobacco manufacturers have used sophisticated techniques to innovate their products in order to reach new consumer subgroups.[54] Strategies fall into two main categories: designing products, especially cigarettes, to have optimal appeal and addictiveness, and communicating the appealing new features to enhance social acceptability. These twin strategies are used in parallel to synergistically augment their impact: Marketing, including pack messaging, reinforces the characteristics of a cigarette that certain consumers find appealing, and the appeal of those characteristics in turn reinforces the perception of the product that marketing communications are promoting. Evidence has shown that tobacco companies have tailored these strategies to address the needs and preferences of subgroups, especially vulnerable populations.[80,81]

As noted earlier, ST was originally designed to appeal to men, especially those of low socioeconomic status and rural background, with robust flavor and high nicotine levels, reinforced by visible and memorable marketing messages.[54] Adolescents have now been targeted with products that make initiation easier, such as variations in menthol and lower nicotine content. The tobacco industry has long claimed that they do not target youth with advertising or promotions, yet there is abundant evidence, including from industry's own internal documents, showing that they do.[82–84] Aggressive advertising conveys images of ST use as rugged, manly, sexually attractive, and cool, while reinforcing social acceptability of use and highlighting the excitement and reward it offers.[85] Youth-oriented marketing was targeted through magazines (e.g., *Rolling Stone*), billboards, and public promotions of sporting events such as motor racing and music festivals.[83] Use was portrayed as masculine, individualistic, risk-taking, and confident.[54] Despite the 1998 Smokeless Tobacco Master Settlement Agreement, evidence shows that adolescents continued to be targeted through youth-oriented magazines.[73]

Other promotional strategies were systematically applied to people of low socioeconomic status, including higher exposure to TV and radio advertising and greater density of tobacco retailers in low-income rural areas.[80] Meanwhile, promotions to change and shape the social acceptability of ST, including promotions of public events and celebrity endorsements, reinforced the message of social acceptance in that demographic. ST ads featured blue collar references; imagery of cowboys, hunters, and racecar drivers projected a rough, resilient, independent masculinity with brands such as Skoal, Grizzly, and Kodiak[86] in magazines such as *Sports Illustrated* and *Popular Mechanics* as well as at point of sale. Magazine messages also had a reinforcing and cumulative impact on social attitudes to smoking, making it more socially acceptable and creating a normative perception that a majority of one's peers smoked and that it was both enjoyable and a useful strategy to relieve stress and other emotional burdens.[83]

Pursuit of new markets: women and smokers

Tobacco manufacturers have a long history of aggressive marketing of products to attain new consumers and retain existing ones. This has certainly been true of ST companies before and after their acquisition by the tobacco giants.

In the past decade, women have been targeted with messages of social acceptability and personal convenience. Snus ads have focused on the convenience of ST in smoke-free environments, the absence of tobacco odor, and lower health risk.[51,52] Moist snuff advertisements have appeared in magazines such as *Glamour*, *Marie Claire*, and *Vogue*, which present snus as attractive and fashionable.[87] The market remains heavily focused on male use, and some evidence suggests that women are unlikely to be easily persuaded, even to reduce the risks of smoking.[88]

Smokers have been targeted with novel products, including snus with the cigarette brand names Camel and Marlboro. This follows a trend among Swedish men, who have used snus to reduce or quit smoking.[89–91] ST is touted in the United States and other developed countries to reduce health concerns and lessen the impact of smoke-free laws, which have made smoking less personally convenient and created an incentive for cessation.[92] Evidence from current smokers around the time of release of these products in the United States suggests that interest in trying reduced exposure products was greatest among women, non-Hispanic whites, and heavy smokers concerned about health risks.[93] However, US smokers have remained generally unenthusiastic about ST in part because of poor taste qualities and inadequate nicotine dosing.[46,94] Most US smokers who have tried snus do not persist; smokers who report current use of snus were likely to say that they were trying to cut down on cigarettes.[94] In contrast, other evidence has shown that, among Swedish men, most snus users do not persist in smoking.[90] Marketing of ST to smokers has now been almost completely supplanted by noncombusted vaping products, which offer a lower-risk alternative that is more socially acceptable and convenient.

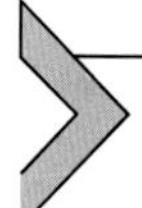

What is the evidence for the impact of innovations on use?

Over the past two decades, tobacco control initiatives in developed countries have lowered demand for conventional products, both combusted and smokeless. This reduction has been driven by price increases, health communication campaigns, pack warnings, cessation support, advertising restrictions, and social denormalization.[95] Innovation in product design and communication underpins the industry's effort to maintain a profitable consumer base in the face of waning demand. Some of the greatest reductions in demand for combusted tobacco have occurred in the United States and other developed countries, and the high perceived health risk of combusted products has been an important contributor to this trend. This has created opportunities for both ST manufacturers, who have sought to expand sales by introducing reduced risk products, and cigarette manufacturers, who have sought to protect dwindling sales by providing ST options for smokers such as Camel Snus and Marlboro Snus.[52] Recently, cigarette manufacturers introduced new options such as Reynolds' Revel lozenges in the United States and pouched products Lyft and Epok in Europe as part of a "modern oral" line.[60] As ST products and marketing evolved from the early 2000s, certain impacts have been seen, including changes in sales and the profile of users. Evidence for the impact of nicotine, flavor, and formulation innovations can be gauged by research on consumer perceptions, prevalence of use, and sales.

Research has shown that overall ST consumption in the United States increased by as much as 23% between 2000 and 2015,[13,96] although other findings suggest that use remained relatively stable from 2002 to 2014.[97,98] However, the broad trend might mask changes in novel product use and population characteristics. For example, the proportion of US adults who were regular users of moist snuff increased 42% between 2001 and 2010, and the proportion of younger adults (aged 18 to 44) who were regular users increased 55%, to 2.8%, corresponding to 2.8 million.[99] Sales of moist snuff, the mainstay of ST manufacturers, increased by 66% from 2005 to 2011.[63] However, sales of newly marketed pouched snus products increased by 334% and contributed 28% of the ST market in the same period. Camel Snus was a top 10 selling moist snuff brand by 2011. Similarly, sales of flavored products increased 72%, contributing 59% of the ST market.[63] These trends were reflected among youth in New York City: While smoking rate declined by 53% from 2001 to 2013, ST rate increased by 400%.[100] During this period, ST quit rate slowed while smoking quit rate increased, further suggesting that ST manufacturers managed to retain market share, perhaps by providing products with lower perceived health risk, which, unlike combusted products, dissuaded consumers from quitting.[101]

While the evidence that might shed light on the impact of innovations in nicotine delivery and formulation in ST products is extremely limited, a growing body of research has addressed the role of flavors in shaping perceptions and patterns of use. Overwhelmingly, flavored tobacco products are perceived more favorably than nonflavored, and are more likely to be used by younger consumers.[102] Qualitative research has shown that flavors are important in promoting appeal among youth[64,103]; and flavors, chiefly mint and wintergreen, are preferred among smokers who switch to ST.[104,105] A study using data from the Population Assessment of Tobacco and Health (PATH) study of 7718 adolescents who had never used tobacco found that susceptibility for ST use was greatest among those who perceived flavored products as easier to use than unflavored.[106] The implication that adolescents will be more likely to use a tobacco product if it is flavored was borne out in several large cross-sectional surveys of US youth and young adults. Over 80% of tobacco users, including combusted forms, e-cigarettes, hookah, and ST, reported using a flavored product.[107,108] Most telling are observations that a majority of users—70% to 81%—initiated with a flavored product,[108,109] and more than 75% of flavored product users reported no interest in continued use of their current tobacco product if it was not flavored.[107] Moreover, users of flavored noncigarette tobacco, which includes ST, have reduced odds of a quit attempt in the past year. While these findings are not all specific to ST, they suggest a strong preference for flavors across all tobacco products among younger users and underscore a reason why manufacturers strive to develop and promote flavored products that appeal to the young. In line with this view are data from the National Adult Tobacco Survey showing that approximately half of an estimated 4 million ST users in the United States used a flavored product in the past 30 days, with use highest among those aged 18 to 24.[105]

Evidence also shows that flavored ST has an advantage in the retail market. Sales of mentholated moist snuff and snus increased from 2011 to 2015, while sales of flavored nonmentholated moist snuff and snus declined.[110] This may be attributable to observations that ST manufacturers have made extensive use of advertising that promotes flavor options, including messages designed to elicit interest in ST among smokers.[54] Moist snuff sales were mostly menthol, which closely reflects findings on consumer preferences: Mentholated brands accounted for 57% of moist snuff sales and 89% of snus.[110] The profound impact of ST flavors on consumer preference and use is further informed by research on tobacco sales in New York City after sale of flavored products was banned in 2009. While flavored tobacco sales declined dramatically as expected,[111,112] further analyses showed that, 3 years after the ban, teens were 37% less likely to ever try a flavored tobacco product and 28% less likely to try any tobacco product.[112]

Thus, recent evidence drawn from a range of sources shows that flavors are a primary driver of interest in use, consumer preference, and current use of ST. Moreover, mint and menthol overwhelmingly are the preferred flavor options. The products sold by ST manufacturers are in close alignment with population use trends: In 2011, 51% of the total ST retail market was accounted for by mint and menthol; this corresponds to a 76% increase in mint and menthol sales from 2005.[63] Manufacturers have sought to meet consumer demand for mint and menthol by designing and marketing those products. As a result, more youth have initiated ST use, and manufacturers have succeeded in expanding overall sales.

However, these broad trends may mask adoption of novel products among nontraditional populations of ST users, including youth, women, and those living in urban settings. For example, analysis of the 2011 US National Youth Tobacco Survey found that, of high school students using ST, 26.8% used a combination of novel and conventional products, while 9.2% used only a novel product.[113] Between 2001 and 2013, use by high school athletes increased by 10%, even while tobacco use was declining among high school nonathletes.[114] Young adults of a sexual minority (gay, lesbian, bisexual, or self-reported as "something else") were 2.1–3.3 times more likely to use ST, based on 2013/14 PATH survey data.[115]

Policy and regulatory strategies for an evolving market

Given the capacity and incentive of manufacturers to refine, innovate, and promote, the potential for expansion of ST use raises clear public health concerns. This is particularly so if their efforts are responsible for initiating use in youth, or if new products dissuade current tobacco users, including smokers, from quitting. Concerns have also been expressed about the potential for dual use of ST and combusted tobacco. While no tobacco use is safe, some forms such as low-nitrosamine ST are likely to lower individual risk compared to smoking.[116] Combusted tobacco use has declined in many developed countries, such as in the United States, where the rate of adult current use is 16.7%.[117] In countries in which consumers are increasingly concerned about the health risks of combusted tobacco, retail markets are rapidly adapting to accommodate consumer preferences. In this changing environment, ST manufacturers are seeking to obtain advantage by innovation, yet they face new challenges as cigarette manufacturers promote vaping and heat-not-burn alternatives.[118] The evidence shows that ST use has not declined at a population level in the United States with the decline of combusted tobacco over the past two decades. That is, ST maintains intrinsic appeal. This means that control measures are needed to prevent the appeal from causing young people to initiate use. At the same time, opportunities to regulate low-risk ST products as alternatives to smoking must be explored. Certainly, challenges posed by ST innovation must be addressed within a comprehensive tobacco control platform, in developed and developing countries alike.

Regulatory approaches to innovation of ST products

The WHO's FCTC, with over 180 countries as parties, has laid out a broad set of evidence-based antitobacco policy strategies with merit for use in regulating ST.[119] The articles call for strategies that include monitoring and surveillance, health warnings and anti-tobacco communications, bans on marketing, taxes, and cessation support. Because the FCTC applies to all tobacco products equally, many if not most policy approaches are directly relevant to ST. Therefore, strategies that include disclosure and regulation of contents (Articles 9 and 10) are especially important for ST, particularly if evidence points to design modifications that increase appeal among youth or mislead consumer perceptions of health risks. Likewise, regulations that restrict marketing, such as advertising or promotions that make claims of lowered risk, are important tools for addressing ST innovation. Even smoke-free laws, a key strategy under the FCTC, are relevant to ST: Because cigarette manufacturers have marketed smokeless products to smokers to help subvert indoor smoking bans,[120] smoke-free laws should be extended to include bans on all tobacco product use in regulated environments.

Regulatory agencies in a number of national jurisdictions have the authority to adopt ST policies. Under an act passed in 2009, the FDA can regulate the manufacture, sale, and marketing of tobacco products.[121] They can require manufacturers to apply for premarket approval of a new product, or for a claim of reduced risk. In October, 2019 the FDA granted the first-ever approval for a modified risk tobacco product claim to General Snus, a Swedish pouch-style snus product manufactured by Swedish Match North America. Similarly, the European Union's Tobacco Products Directive of 2014 regulates manufacture, presentation, and sale in member states. It bans the sale and marketing of Swedish snus, with exceptions for Sweden and Norway, where snus has a long-standing history.[122] Other forms of ST with a limited customer base, such as chewing tobacco and nasal snuff, are not subject to the directive.[123] Bans on ST sale and marketing have been introduced in a few other countries: Australia, Hong Kong, Singapore, and the United Arab Emirates. In India, the Food Safety and Standards Authority banned the sale of gutkha following a Supreme Court decision in 2012 pertaining to food safety laws.[124] However, product innovations have been used to subvert this regulation: Manufacturers of gutkha have since marketed and sold tobacco and other key constituents separately in "twin packs," which are then combined by the consumer.[125]

Globally, there are relatively few examples of regulatory actions that affect innovation in ST products to attract consumers. Rather, much of the focus has been on restriction of flavor additives, notably menthol. Flavored ST has been banned in a few countries, including Ethiopia, Chile, and Moldova. Canada banned flavor additives in cigarettes in 2009 and extended the ban to menthol in 2017. ST, waterpipe (shisha) tobacco, and alcohol-flavored cigarillos are not covered, leaving a potential regulatory loophole.[126] Likewise, the European Union banned flavored cigarettes in 2016, and Turkey banned menthol cigarettes in 2015, though these bans do not extend to other tobacco products. Brazilian regulators approved a ban on flavor additives, including menthol, in all tobacco products in 2012, yet the regulation remains suspended by litigation. In the United States, the FDA issued an advance notice of proposed rulemaking (ANPRM) in 2018 that will consider regulatory approaches for flavors in noncigarette tobacco products, including ST. The ANPRM will assess the role of flavors in youth initiation, as well as whether flavors play a role in helping adult smokers switch to a lower risk ST product.[127] Similarly, the FDA is considering regulating menthol in combusted and noncombusted products, which could see the future adoption of regulations on menthol in ST.[128] Globally, flavor regulations are complex, often beset with loopholes, vary across national and subnational jurisdictions, and will likely continue to change over time. Effective regulations on flavored ST should seek to prevent manufacturers from developing products that appeal to youth, but should also consider whether the products might encourage adult smokers to switch.

Regulation of nicotine level or the form of nicotine delivered in ST may help reduce abuse liability, yet this approach has seldom been proposed. While the FDA has issued an ANPRM to reduce nicotine below an addictive threshold, this applies only to cigarettes.[129] Few if any other countries have proposed regulation of nicotine, and ST remains an underrecognized regulatory target. Similarly, very few jurisdictions have sought to regulate presentation and packaging of ST, including pack size or, for pouched products such as snus, the size of the pouch. In South Africa, snus may be sold in cans of various sizes, including small ones which are more attractive to people unable to afford larger ones. In the United States, cigarettes have a mandated minimum pack size of 20, yet snus does not have a minimum number of pouches per pack. Smaller volume packs, which are lower priced, are more appealing and accessible to youth.[130]

Many jurisdictions have broad regulations to limit tobacco product advertising and promotion, and these generally apply to ST. Still, while articles of the FCTC apply equally to all tobacco products, they have not been evenly applied to ST and are often overlooked by the membership. A recent analysis found that just 16 (9%) of parties have adopted a ban on ST advertisement, promotion, and sponsorships.[17] This is concerning in light of the rapidly evolving nature of digital technology and the opportunities to market products using websites, apps, and social media. In the United States, the FDA has reissued a rule restricting sponsorship of sporting, entertainment, and social or cultural events by tobacco manufacturers, including ST. However, the current rules do not apply to digital media, leaving the door open for novel promotion technique and products.[131,132] Regulations that impose restrictions on product pack descriptors, colors, and other features have been imposed in some jurisdictions for combusted tobacco, but rarely for ST. Evidence-based regulation on the promotion of ST innovations via mass communication such as digital media and pack- and point-of-sale methods is urgently needed.[132]

Summary and conclusions

The rapidly changing tobacco market in some developed countries has seen a shift in consumer preferences from combusted toward noncombusted products. Manufacturers that traditionally made cigarettes are becoming increasingly involved in ST. As noncombusted forms of nicotine delivery become normalized, new market dynamics will come into play. ST will face competition from e-cigarettes, and evidence-based tobacco control interventions will continue to restrict ST, so that innovations in product design and marketing will be increasingly important. A nuanced understanding of innovation and its impact on retaining and expanding markets, especially those in developing countries, is necessary to protect public health. At the same time, urgent actions are needed in regions where the greatest proportion of ST users are located, even though innovation has occurred at a slower pace there. For example, in India, where some 90% of the world's ST users live, policies can be applied at state and local levels to reduce product appeal.[133,134] At a minimum, all parties to the FCTC should adopt and enforce a full complement of policies that aim to reduce ST demand.[17]

Because ST poses lower health risks than combusted products, regulatory opportunities to optimize harm reduction should be considered.[135,136] ST should be regulated in a way that supports a net reduction in tobacco-related harm. Such policies should serve to prevent youth initiation, and policymakers must rigorously evaluate and regulate health claims made by ST manufacturers. Consumer responses to FDA-approved claims for modified risk, such as granted to General Snus, should be monitored to determine whether the use of modified risk products leads not only to lowered exposure but also to reduced consumption of products with higher health risks. ST product regulations should be aligned with those for combusted products so that lower risk products are available to adult smokers while rigorous measures are applied to prevent youth initiation.

Global lessons suggest that regulation should seek to limit the proliferation of new ST markets and to lower use in current markets by blocking opportunities to innovate products. However, relatively few countries have implemented regulations on ST products that would restrict manufacturers' ability to innovate and recruit new consumers. The FCTC and other regulatory mechanisms, including the US FSPTC Act and the European Directive on Tobacco Products, provide promising opportunities. Standards that require the elimination of toxic constituents to the extent possible should be adopted. Standards that restrict flavors and regulations on packaging and mass marketing strategies, including digital media, will minimize the appeal to youth. Standards that regulate nicotine delivery and the ease and convenience of use must be developed to support adult smokers' ability to switch to ST. An ambitious research agenda is needed to support standards that will yield minimal-risk alternatives to combusted tobacco. Existing comprehensive approaches to ST regulation, which include increased taxes, health warnings, cessation support, and antitobacco health communications, are effective and will continue to be needed.[137] In the context of a changing tobacco market, policies that limit ST manufacturers' capacity to develop new global markets while providing options for smokers to reduce their risk from more deadly combusted products will be indispensable.

References

1. IARC Working Group on the Evaluation of Carcinogenic Risk to Humans. *Personal Habits and Indoor Combustions. Smokeless Tobacco. IARC Monographs on the Evaluation of Carcinogenic Risks to Humans, No. 100E*. Lyon: International Agency for Research on Cancer; 2012.
2. Lee K, Eckhardt J. The globalisation strategies of five Asian tobacco companies: a comparative analysis and implications for global health governance. *Glob Public Health*. 2017;12(3):367–379. https://doi.org/10.1080/17441692.2016.1273370.
3. Lewis MJ, Wackowski O. Dealing with an innovative industry: a look at flavored cigarettes promoted by mainstream brands. *Am J Public Health*. 2006;96(2):244–251. https://doi.org/10.2105/AJPH.2004.061200.

4. WHO Framework Convention on Tobacco Control. *Commonly Used Smokeless Tobacco Products Around the Globe*; 2019. Geneva https://untobaccocontrol.org/kh/smokeless-tobacco/paan-betel-quid-tobacco/.
5. National Cancer Institute and Centers for Disease Control and Prevention. *Smokeless Tobacco and Public Health: A Global Perspective*. Bethesda, MD: U.S. Department of Health and Human Services, Centers for Disease Control and Prevention and National Institutes of Health, National Cancer Institute. NIH Publication No. 14-7983; 2014.
6. Cullen D, Keithly L, Kane K, et al. Smokeless tobacco products sold in Massachusetts from 2003 to 2012: trends and variations in brand availability, nicotine contents and design features. *Tob Control*. 2015;24(3):256–262. https://doi.org/10.1136/tobaccocontrol-2013-051225.
7. Mehrotra R, Sinha DN, Tibor S. *Global Smokeless Tobacco Control Policies and Their Implementation*. Uttar Pradesh: National Institute of Cancer Prevention and Research; 2017.
8. Guven A, Tolun F. Effects of smokeless tobacco "maras powder" use on nitric oxide and cardiovascular risk parameters. *Int J Med Sci*. 2012;9(9):786–792. https://doi.org/10.7150/ijms.4563.
9. WHO South East Asian Regional Office (SEARO). *Report on Oral Tobacco Use and Its Implications in South-East Asia*; 2004. New Delhi http://www.searo.who.int/tobacco/topics/oral_tobacco_use.pdf.
10. Bhisey RA. Chemistry and toxicology of smokeless tobacco. *Indian J Cancer*. 2012;49(4):364–372. https://doi.org/10.4103/0019-509X.107735.
11. Euromonitor International. *Passport: Tobacco. Market Sizes*. 2019. Chicago, IL.
12. Singh A, Ladusingh L. Prevalence and determinants of tobacco use in India: evidence from recent Global Adult Tobacco Survey dataGorlova OY, ed. *PLoS One*. 2014;9(12):e114073. https://doi.org/10.1371/journal.pone.0114073.
13. Agaku IT, Alpert HR. Trends in annual sales and current use of cigarettes, cigars, roll-your-own tobacco, pipes, and smokeless tobacco among US adults, 2002–2012. *Tob Control*. 2016;25(4):451–457. https://doi.org/10.1136/tobaccocontrol-2014-052125.
14. Federal Trade Commission. *Federal Trade Commission Smokeless Tobacco Report for 2017*. 2019. Washington, DC.
15. Mohan P, Lando HA, Panneer S. Assessment of tobacco consumption and control in India. *Indian J Clin Med*. 2018;9. https://doi.org/10.1177/1179916118759289.
16. World Health Organization. *90% of Smokeless Tobacco Users Live in South-East Asia*; 2016. http://www.searo.who.int/mediacentre/releases/2013/pr1563/en/.
17. Mehrotra R, Yadav A, Sinha DN, et al. Smokeless tobacco control in 180 countries across the globe: call to action for full implementation of WHO FCTC measures. *Lancet Oncol*. 2019;20(4):e208–e217. https://doi.org/10.1016/S1470-2045(19)30084-1.
18. Macy MJT, Li J, Xun P, Presson CC, Chassin L. Dual trajectories of cigarette smoking and smokeless tobacco use from adolescence to midlife among males in a midwestern US community sample. *Nicotine Tob Res*. 2016;18(2):186–195. https://doi.org/10.1093/ntr/ntv070.
19. Jasinski DR, Henningfield JE. Human abuse liability assessment by measurement of subjective and physiological effects. *NIDA Res Monogr*. 1989;92:73–100.
20. Carter LP, Griffiths RR. Principles of laboratory assessment of drug abuse liability and implications for clinical development. *Drug Alcohol Depend*. 2009;105(Suppl):S14–S25.
21. Hoffmann D, Djordjevic MV, Hoffmann I. The changing cigarette. *Prev Med*. 1997;26(4):427–434. https://doi.org/10.1006/pmed.1997.0183.

22. Alpert HR, Koh H, Connolly GN. Free nicotine content and strategic marketing of moist snuff tobacco products in the United States: 2000–2006. *Tob Control*. 2008; 17(5):332–338. https://doi.org/10.1136/tc.2008.025247.
23. Duprey R. *Latest Altria Acquisition Bolsters its Smokeless Segment*. Yahoo Finance; June 2019. Published https://finance.yahoo.com/news/latest-altria-acquisition-bolsters-smokeless-170900208.html.
24. JUUL Labs. JUUL Statement About Altria Minority Investment and Service Agreements. https://ryghub.com/2018/12/juul-labs-issues-statement-about-altria-minority-investment-and-service-agreements/.
25. U.S. Food and Drug Administration. *FDA Permits Sale of IQOS Tobacco Heating System through Premarket Tobacco Product Application Pathway*; 2019. Washington, D.C. https://www.fda.gov/news-events/press-announcements/fda-permits-sale-iqos-tobacco-heating-system-through-premarket-tobacco-product-application-pathway
26. Kozlowski LT, Abrams DB. Obsolete tobacco control themes can be hazardous to public health: the need for updating views on absolute product risks and harm reduction. *BMC Public Health*. 2016;16(1):432. https://doi.org/10.1186/s12889-016-3079-9.
27. Berg CJ, Haardörfer R, Getachew B, Johnston T, Foster B, Windle M. Fighting fire with fire. *Soc Mark Q*. 2017;23(4):302–319. https://doi.org/10.1177/1524500417718533.
28. Lisha NE, Jordan JW, Ling PM. Peer crowd affiliation as a segmentation tool for young adult tobacco use. *Tob Control*. 2016;25(Suppl 1):i83–i89. https://doi.org/10.1136/tobaccocontrol-2016-053086.
29. Kasza KA, Ambrose BK, Conway KP, et al. Tobacco product use by adults and youths in the United States in 2013 and 2014. *N Engl J Med*. 2017;376(4):342–353. https://doi.org/10.1056/NEJMsa1607538.
30. O'Connor RJ, Cummings KM, Rees VW, et al. Surveillance methods for identifying, characterizing, and monitoring tobacco products: potential reduced exposure products as an example. *Cancer Epidemiol Biomarkers Prev*. 2009;18(12):3334–3348. https://doi.org/10.1158/1055-9965.EPI-09-0429.
31. Wayne GF, Connolly GN. How cigarette design can affect youth initiation into smoking: Camel cigarettes 1983–93. *Tob Control*. 2002;11(suppl 1):I32–I39. https://doi.org/10.1136/TC.11.SUPPL_1.I32.
32. Carpenter CM, Wayne GF, Connolly GN. The role of sensory perception in the development and targeting of tobacco products. *Addiction*. 2007;102(1):136–147.
33. Kreslake JM, Wayne GF, Alpert HR, Koh HK, Connolly GN. Tobacco industry control of menthol in cigarettes and targeting of adolescents and young adults. *Am J Public Health*. 2008;98(9):1685–1692.
34. Rees VW, Kreslake JM, Cummings KM, et al. Assessing consumer responses to potential reduced-exposure tobacco products: a review of tobacco industry and independent research methods. *Cancer Epidemiol Biomark Prev*. 2009;18(12). https://doi.org/10.1158/1055-9965.EPI-09-0946.
35. Carter LP, Stitzer ML, Henningfield JE, O'Connor RJ, Cummings KM, Hatsukami DK. Abuse liability assessment of tobacco products including potential reduced exposure products. *Cancer Epidemiol Biomark Prev*. 2009;18(12):3241–3262. https://doi.org/10.1158/1055-9965.EPI-09-0948.
36. Balfour DJK. The neuronal pathways mediating the behavioral and addictive properties of nicotine. In: *Nicotine Psychopharmacology*. Berlin, Heidelberg: Springer Berlin Heidelberg; 2009:209–233. https://doi.org/10.1007/978-3-540-69248-5_8.
37. Fant RV, Henningfield JE, Nelson RA, Pickworth WB. Pharmacokinetics and pharmacodynamics of moist snuff in humans. *Tob Control*. 1999;8(4):387–392. https://doi.org/10.1136/tc.8.4.387.

38. Tomar SL, Henningfield JE. Review of the evidence that pH is a determinant of nicotine dosage from oral use of smokeless tobacco. *Tob Control.* 1997;6(3):219–225. http://www.ncbi.nlm.nih.gov/pubmed/9396107.
39. Centers for Disease Control and Prevention (CDC). Determination of nicotine, pH, and moisture content of six U.S. commercial moist snuff products–Florida, January-February 1999. *MMWR Morb Mortal Wkly Rep.* 1999;48(19):398–401. http://www.ncbi.nlm.nih.gov/pubmed/10366135.
40. B Pickworth W, Rosenberry ZR, Gold W, Koszowski B. Nicotine absorption from smokeless tobacco modified to adjust pH. *J Addict Res Ther.* 2014;05(03):1–5. https://doi.org/10.4172/2155-6105.1000184.
41. Ferris Wayne G, Connolly GN, Henningfield JE. Brand differences of free-base nicotine delivery in cigarette smoke: the view of the tobacco industry documents. *Tob Control.* 2006;15(3):189–198. https://doi.org/10.1136/tc.2005.013805.
42. Tilashalski K, Rodu B, Mayfield C. Assessing the nicotine content of smokeless tobacco products. *J Am Dent Assoc.* 1994;125(5), 590-592, 594 http://www.ncbi.nlm.nih.gov/pubmed/8195501.
43. Henningfield JE, Radzius A, Cone EJ. Estimation of available nicotine content of six smokeless tobacco products. *Tob Control.* 1995;4(1):57. https://www.ncbi.nlm.nih.gov/pmc/articles/PMC1759397/.
44. Tomar SL, Giovino GA, Eriksen MP. Smokeless tobacco brand preference and brand switching among US adolescents and young adults. *Tob Control.* 1995;4:67–72. https://doi.org/10.2307/20747348.
45. Borgerding MF, Bodnar JA, Curtin GM, Swauger JE. The chemical composition of smokeless tobacco: a survey of products sold in the United States in 2006 and 2007. *Regul Toxicol Pharmacol.* 2012;64(3):367–387. https://doi.org/10.1016/j.yrtph.2012.09.003.
46. Seidenberg AB, Ayo-Yusuf OA, Rees VW. Characteristics of "American snus" and Swedish snus products for sale in Massachusetts, USA. *Nicotine Tob Res.* 2018;20(2). https://doi.org/10.1093/ntr/ntw334.
47. Feirman SP, Donaldson EA, Parascandola M, Snyder K, Tworek C. Monitoring harm perceptions of smokeless tobacco products among U.S. adults: health Information National Trends Survey 2012, 2014, 2015. *Addict Behav.* 2018;77:7–15. https://doi.org/10.1016/j.addbeh.2017.09.002.
48. Rutqvist LE, Curvall M, Hassler T, Ringberger T, Wahlberg I. Swedish snus and the GothiaTek® standard. *Harm Reduct J.* 2011;8(1):11. https://doi.org/10.1186/1477-7517-8-11.
49. Rickert WS, Joza PJ, Trivedi AH, Momin RA, Wagstaff WG, Lauterbach JH. Chemical and toxicological characterization of commercial smokeless tobacco products available on the Canadian market. *Regul Toxicol Pharmacol.* 2009;53(2):121–133. https://doi.org/10.1016/j.yrtph.2008.12.004.
50. Ayo-Yusuf OA, Connolly GN. Applying toxicological risk assessment principles to constituents of smokeless tobacco products: implications for product regulation. *Tob Control.* 2011;20(1):53–57. https://doi.org/10.1136/tc.2010.037135.
51. Sami M, Timberlake DS, Nelson R, et al. Smokers' perceptions of smokeless tobacco and harm reduction. *J Public Health Policy.* 2012;33(2):188–201. https://doi.org/10.1057/jphp.2012.9.
52. Timberlake DS, Pechmann C, Tran SY, Au V. A content analysis of Camel Snus advertisements in print media. *Nicotine Tob Res.* 2011;13(6):431–439. https://doi.org/10.1093/ntr/ntr020.
53. Agaku IT, Ayo-Yusuf OA. The effect of exposure to pro-tobacco advertising on experimentation with emerging tobacco products among U.S. adolescents. *Health Educ Behav.* 2014;41(3):275–280. https://doi.org/10.1177/1090198113511817.

54. Mejia AB, Ling PM. Tobacco industry consumer research on smokeless tobacco users and product development. *Am J Public Health*. 2010;100(1):78–87. https://doi.org/10.2105/AJPH.2008.152603.
55. Connolly GN, Richter P, Aleguas A, Pechacek TF, Stanfill SB, Alpert HR. Unintentional child poisonings through ingestion of conventional and novel tobacco products. *Pediatrics*. 2010;125(5):896–899. https://doi.org/10.1542/peds.2009-2835.
56. Wang B, Rostron B. Tobacco-related poison events involving young children in the US, 2001–2016. *Tob Regul Sci*. 2017;3(4):525–535. https://doi.org/10.18001/TRS.3.4.12.
57. Tobacco Products Scientific Advisory Committee. *TPSAC Report on Dissolvable Tobacco Products*; 2012. Washington, D.C.; https://wayback.archive-it.org/7993/20170405201701/https://www.fda.gov/downloads/AdvisoryCommittees/CommitteesMeetingMaterials/TobaccoProductsScientificAdvisoryCommittee/UCM295842.pdf.
58. CSP Daily News. *Star Scientific Exiting Tobacco Business*. CSP Daily News; 2012. https://www.cspdailynews.com/tobacco/star-scientific-exiting-tobacco-business.
59. Wall Street Journal. R.J. Reynolds Scales Back Marketing of Dissolvable Tobacco Products | Business | journalnow.com. https://www.journalnow.com/business/business_news/local/r-j-reynolds-scales-back-marketing-of-dissolvable-tobacco-products/article_9d001b58-f9f2-11e2-8fad-0019bb30f31a.html.
60. Carver R. Reynolds expands reach of dissolvable tobacco product with age-21 restrictions. *Winston Salem J*; March 30, 2019. https://www.journalnow.com/business/reynolds-expands-reach-of-dissolvable-tobacco-product-with-age-/article_ded61d79-cb05-525c-9188-26d8e46c796d.html.
61. Brown JE, Luo W, Isabelle LM, Pankow JF. Candy flavorings in tobacco. *N Engl J Med*. 2014;370(23):2250–2252. https://doi.org/10.1056/NEJMc1403015.
62. Tobacco Control Legal Consortium. *Flavored Tobacco on Sale in Minnesota: Research Findings*. Minnesota; 2013. https://www.publichealthlawcenter.org/sites/default/files/resources/tclc-fs-Morris-MN-flavored-tobacco-sold-2013.pdf.
63. Delnevo CD, Wackowski OA, Giovenco DP, Manderski MTB, Hrywna M, Ling PM. Examining market trends in the United States smokeless tobacco use: 2005–2011. *Tob Control*. 2014;23(2):107–112. https://doi.org/10.1136/tobaccocontrol-2012-050739.
64. Kostygina G, Ling PM. Tobacco industry use of flavourings to promote smokeless tobacco products. *Tob Control*. 2016;25(Suppl 2):ii40–ii49. https://doi.org/10.1136/tobaccocontrol-2016-053212.
65. Rainey CL, Conder PA, Goodpaster JV. Chemical characterization of dissolvable tobacco products promoted to reduce harm. *J Agric Food Chem*. 2011;59(6):2745–2751. https://doi.org/10.1021/jf103295d.
66. Mejia AB, Ling PM, Glantz SA. Quantifying the effects of promoting smokeless tobacco as a harm reduction strategy in the USA. *Tob Control*. 2010;19(4):297–305. https://doi.org/10.1136/tc.2009.031427.
67. Bansal-Travers M, O'Connor R, Fix BV, Cummings KM. What do cigarette pack colors communicate to smokers in the U.S.? *Am J Prev Med*. 2011;40(6):683–689. https://doi.org/10.1016/j.amepre.2011.01.019.
68. Connolly GN, Alpert HR. Has the tobacco industry evaded the FDA's ban on 'Light' cigarette descriptors? *Tob Control*. 2014;23(2):140–145. https://doi.org/10.1136/TOBACCOCONTROL-2012-050746.
69. Adkison SE, Bansal-Travers M, Smith DM, O'Connor RJ, Hyland AJ. Impact of smokeless tobacco packaging on perceptions and beliefs among youth, young adults, and adults in the U.S: findings from an internet-based cross-sectional survey. *Harm Reduct J*. 2014;11(1):2. https://doi.org/10.1186/1477-7517-11-2.

70. Ernster VL. Advertising and promotion of smokeless tobacco products. *NCI (Natl Cancer Inst) Monogr.* 1989;8:87–94. http://www.ncbi.nlm.nih.gov/pubmed/2654652.
71. Lynch BS, Bonnie RJ, Institute of Medicine (U.S.), Committee on Preventing Nicotine Addiction in Children and Youths. *Growing up Tobacco Free : Preventing Nicotine Addiction in Children and Youths.* National Academy Press; 1994.
72. National Association of Attorneys General. *Smokeless Tobacco Master Settlement Agreement*; 1998. Washington, D.C https://www.naag.org/assets/redesign/files/msa-tobacco/STMSA.pdf.
73. Morrison MA, Krugman DM, Park P. Under the radar: smokeless tobacco advertising in magazines with substantial youth readership. *Am J Public Health.* 2008;98(3):543–548. https://doi.org/10.2105/AJPH.2006.092775.
74. Moran MB, Heley K, Baldwin K, Xiao C, Lin V, Pierce JP. Selling tobacco: a comprehensive analysis of the U.S. tobacco advertising landscape. *Addict Behav.* 2019;96:100–109. https://doi.org/10.1016/j.addbeh.2019.04.024.
75. O'Brien EK, Navarro MA, Hoffman L. Mobile website characteristics of leading tobacco product brands: cigarettes, smokeless tobacco, e-cigarettes, hookah and cigars. *Tob Control.* August 2018. https://doi.org/10.1136/tobaccocontrol-2018-054549.
76. Seidenberg AB, Rees VW, Connolly GN. Swedish Match marketing on YouTube. *Tob Control.* 2010;19(6):512–513. https://doi.org/10.1136/tc.2010.038919.
77. Bromberg JE, Augustson EM, Backinger CL. Portrayal of smokeless tobacco in YouTube videos. *Nicotine Tob Res.* 2012;14(4):455–462. https://doi.org/10.1093/ntr/ntr235.
78. Navarro MA, O'Brien EK, Hoffman L. Cigarette and smokeless tobacco company smartphone applications. *Tob Control.* July 2018. https://doi.org/10.1136/tobaccocontrol-2018-054480.
79. Seidenberg AB, Rodgers EJ, Rees VW, Connolly GN. Youth access, creation, and content of smokeless tobacco ("dip") videos in social media. *J Adolesc Health.* 2012;50(4):334–338. https://doi.org/10.1016/j.jadohealth.2011.09.003.
80. Hackbarth DP, Silvestri B, Cosper W. Tobacco and alcohol billboards in 50 Chicago neighborhoods: market segmentation to sell dangerous products to the poor. *J Public Health Policy.* 1995;16(2):213. https://doi.org/10.2307/3342593.
81. Seidenberg AB, Caughey RW, Rees VW, Connolly GN. Storefront cigarette advertising differs by community demographic profile. *Am J Health Promot.* 2010;24(6):e26–31. https://doi.org/10.4278/ajhp.090618-QUAN-196.
82. U.S. Department of Health and Human Services. *Preventing Tobacco Use Among Youth and Young Adults : A Report of the Surgeon General.* Atlanta, GA: Dept. of Health and Human Services, Centers for Disease Control and Prevention, National Center for Chronic Disease Prevention and Health Promotion, Office on Smoking and Health; 2012.
83. National Cancer Institute. *The Role of the Media in Promoting and Reducing Tobacco use. Tobacco Control Monograph No 19.* Bethesda, MD. 2008.
84. Cummings KM, Morley CP, Horan JK, Steger C, Leavell N-R. Marketing to America's youth: evidence from corporate documents. *Tob Control.* 2002;11(Suppl 1):I5–I17. http://www.ncbi.nlm.nih.gov/pubmed/11893810.
85. Kostygina G, Glantz SA, Ling PM. Tobacco industry use of flavours to recruit new users of little cigars and cigarillos. *Tob Control.* 2014;25(1). https://doi.org/10.1136/tobaccocontrol-2014-051830.
86. Hendlin YH, Veffer JR, Lewis MJ, Ling PM. Beyond the brotherhood: Skoal Bandits' role in the evolution of marketing moist smokeless tobacco pouches. *Tob Induc Dis.* 2017;15(1):46. https://doi.org/10.1186/s12971-017-0150-y.

87. Dave D, Saffer H. Demand for smokeless tobacco: role of advertising. *J Health Econ.* 2013;32(4):682–697. https://doi.org/10.1016/j.jhealeco.2013.03.007.
88. Popova L, Kostygina G, Sheon NM, Ling PM. A qualitative study of smokers' responses to messages discouraging dual tobacco product use. *Health Educ Res.* 2014; 29(2):206–221. https://doi.org/10.1093/her/cyt150.
89. Lund KE, Scheffels J, McNeill A. The association between use of snus and quit rates for smoking: results from seven Norwegian cross-sectional studies. *Addiction.* 2011;106(1): 162–167. https://doi.org/10.1111/j.1360-0443.2010.03122.x.
90. Ramström L, Borland R, Wikmans T. Patterns of smoking and snus use in Sweden: implications for public health. *Int J Environ Res Public Health.* 2016;13(11):1110. https://doi.org/10.3390/ijerph13111110.
91. Maki J. The incentives created by a harm reduction approach to smoking cessation: snus and smoking in Sweden and Finland. *Int J Drug Policy.* 2015;26(6):569–574. https://doi.org/10.1016/j.drugpo.2014.08.003.
92. Pederson LL, Nelson DE. Literature review and summary of perceptions, attitudes, beliefs, and marketing of potentially reduced exposure products: communication implications. *Nicotine Tob Res.* 2007;9(5):525–534. https://doi.org/10.1080/14622200701239548.
93. Parascandola M, Augustson E, O'Connell ME, Marcus S. Consumer awareness and attitudes related to new potential reduced-exposure tobacco product brands. *Nicotine Tob Res.* 2009;11(7):886–895. https://doi.org/10.1093/ntr/ntp082.
94. Biener L, Roman AM, Inerney SAM, et al. Snus use and rejection in the USA. *Tob Control.* 2016;25(4):386–392. https://doi.org/10.1136/TOBACCOCONTROL-2013-051342.
95. U.S. Department of Health and Human Services. *The Health Consequences of Smoking—50 Years of Progress: A Report of the Surgeon General.* 2014. Atlanta, GA.
96. Wang TW, Kenemer B, Tynan MA, Singh T, King B. Consumption of combustible and smokeless tobacco - United States, 2000–2015. *MMWR Morb Mortal Wkly Rep.* 2016;65(48):1357–1363. https://doi.org/10.15585/mmwr.mm6548a1.
97. Lipari RN, Van Horn SL. *Trends in Smokeless Tobacco Use and Initiation: 2002 to 2014*; 2013. http://www.ncbi.nlm.nih.gov/pubmed/28636307.
98. Chang JT, Levy DT, Meza R. Trends and factors related to smokeless tobacco use in the United States. *Nicotine Tob Res.* 2016;18(8):1740–1748. https://doi.org/10.1093/ntr/ntw090.
99. Bhattacharyya N. Trends in the use of smokeless tobacco in United States, 2000–2010. *The Laryngoscope.* 2012;122(10):2175–2178. https://doi.org/10.1002/lary.23448.
100. Elfassy T, Yi SS, Kansagra SM. Trends in cigarette, cigar, and smokeless tobacco use among New York City public high school youth smokers, 2001–2013. *Prev Med Rep.* 2015;2:488–491. https://doi.org/10.1016/j.pmedr.2015.06.009.
101. Chang JT, Levy DT, Meza R. Examining the transitions between cigarette and smokeless tobacco product use in the United States using the 2002–2003 and 2010–2011 longitudinal cohorts. *Nicotine Tob Res.* 2018;20(11):1412–1416. https://doi.org/10.1093/ntr/ntx251.
102. Feirman SP, Lock D, Cohen JE, Holtgrave DR, Li T. Flavored tobacco products in the United States: a systematic review assessing use and attitudes. *Nicotine Tob Res.* 2016;18(5):739–749. https://doi.org/10.1093/ntr/ntv176.
103. Couch ET, Darius EF, Walsh MM, Chaffee BW. ST product characteristics and relationships with perceptions and behaviors among rural adolescent males: a qualitative study. *Health Educ Res.* 2017;32(6):537–545. https://doi.org/10.1093/her/cyx067.

104. Oliver AJ, Jensen JA, Vogel RI, Anderson AJ, Hatsukami DK. Flavored and nonflavored smokeless tobacco products: rate, pattern of use, and effects. *Nicotine Tob Res.* 2013;15(1):88. https://doi.org/10.1093/NTR/NTS093.
105. Bonhomme MG, Holder-Hayes E, Ambrose BK, et al. Flavoured non-cigarette tobacco product use among US adults: 2013–2014. *Tob Control.* 2016;25(Suppl 2):ii4–ii13. https://doi.org/10.1136/tobaccocontrol-2016-053373.
106. Chaffee BW, Urata J, Couch ET, Gansky SA. Perceived flavored smokeless tobacco ease-of-use and youth susceptibility. *Tob Regul Sci.* 2017;3(3):367–373. https://doi.org/10.18001/TRS.3.3.12.
107. Harrell MB, Loukas A, Jackson CD, Marti CN, Perry CL. Flavored tobacco product use among youth and young adults: what if flavors didn't exist? *Tob Regul Sci.* 2017; 3(2):168–173. https://doi.org/10.18001/TRS.3.2.4.
108. Villanti AC, Johnson AL, Ambrose BK, et al. Flavored tobacco product use in youth and adults: findings from the first wave of the PATH study (2013–2014). *Am J Prev Med.* 2017;53(2):139–151. https://doi.org/10.1016/j.amepre.2017.01.026.
109. Smith DM, Bansal-Travers M, Huang J, Barker D, Hyland AJ, Chaloupka F. Association between use of flavoured tobacco products and quit behaviours: findings from a cross-sectional survey of US adult tobacco users. *Tob Control.* 2016;25(Suppl 2):ii73–ii80. https://doi.org/10.1136/tobaccocontrol-2016-053313.
110. Kuiper NM, Gammon D, Loomis B, et al. Trends in sales of flavored and menthol tobacco products in the United States during 2011–2015. *Nicotine Tob Res.* 2018; 20(6):698–706. https://doi.org/10.1093/ntr/ntx123.
111. Rogers T, Brown EM, McCrae TM, et al. Compliance with a sales policy on flavored non-cigarette tobacco products. *Tob Regul Sci.* 2017;3(2 Suppl 1):S84–S93. https://doi.org/10.18001/TRS.3.2(Suppl1).9.
112. Farley SM, Johns M. New York City flavoured tobacco product sales ban evaluation. *Tob Control.* 2017;26(1):78–84. https://doi.org/10.1136/tobaccocontrol-2015-052418.
113. Agaku IT, Ayo-Yusuf OA, Vardavas CI, Alpert HR, Connolly GN. Use of conventional and novel smokeless tobacco products among US adolescents. *Pediatrics.* 2013; 132(3):e578–e586. https://doi.org/10.1542/peds.2013-0843.
114. Agaku IT, Singh T, Jones SE, et al. Combustible and smokeless tobacco use among high school athletes — United States, 2001–2013. *MMWR Morb Mortal Wkly Rep.* 2015;64(34):935–939. https://doi.org/10.15585/mmwr.mm6434a2.
115. Wheldon CW, Kaufman AR, Kasza KA, Moser RP. Tobacco use among adults by sexual orientation: findings from the population assessment of tobacco and health study. *LGBT Health.* 2018;5(1):33–44. https://doi.org/10.1089/lgbt.2017.0175.
116. Levy DT, Mumford EA, Cummings KM, et al. The relative risks of a low-nitrosamine smokeless tobacco product compared with smoking cigarettes: estimates of a panel of experts. *Cancer Epidemiol Biomark Prev.* 2004;13(12):2035–2042. http://www.ncbi.nlm.nih.gov/pubmed/15598758.
117. Wang TW, Asman K, Gentzke AS, et al. Tobacco product use among adults — United States, 2017. *MMWR Morb Mortal Wkly Rep.* 2018;67(44):1225–1232. https://doi.org/10.15585/mmwr.mm6744a2.
118. Stone E, Marshall H. Tobacco and electronic nicotine delivery systems regulation. *Transl Lung Cancer Res.* 2019;8(Suppl 1):S67–S76. https://doi.org/10.21037/tlcr.2019.03.13.
119. WHO. Framework Convention on Tobacco Control. http://www.who.int/fctc/en/.
120. Carpenter CM, Connolly GN, Ayo-Yusuf OA, Wayne GF. Developing smokeless tobacco products for smokers: an examination of tobacco industry documents. *Tob Control.* 2009;18(1):54–59. https://doi.org/10.1136/tc.2008.026583.
121. US Department of Health and Human Services Food and Drug Administration. *Family Smoking Prevention and Tobacco Control Act.* Public Law; 2009:111–131.

122. Directive 2014/40/EU of the European Parliament and of the Council of 3 April 2014 on the Approximation of the Laws, Regulations and Administrative Provisions of the Member States Concerning the Manufacture, Presentation and Sale of Tobacco and Related Pr.
123. Leon ME, Lugo A, Boffetta P, et al. Smokeless tobacco use in Sweden and other 17 European countries. *Eur J Public Health.* 2016;26(5):817–821. https://doi.org/10.1093/eurpub/ckw032.
124. Yadav A, Singh A, Khadka BB, Amarasinghe H, Yadav N, Singh R. Smokeless tobacco control: litigation & judicial measures from Southeast Asia. *Indian J Med Res.* 2018;148(1):25–34. https://doi.org/10.4103/ijmr.IJMR_2063_17.
125. Kumar G, Kumar P. Smokeless tobacco (SLT) use in Delhi after three years of ban on gutka and one Year on all SLT products. *Natl J Commun Med.* 2018;9(11):836–839.
126. Government of Canada. Order amending the schedule to the tobacco act (menthol) P.C. 2017-256 March 24, 2017. *Canada Gaz.* 2017;151(7).
127. US Department of Health and Human Services Food and Drug Administration. Regulation of flavors in tobacco products. 21 CFR parts 1100, 1140, and 1143 [Docket No. FDA–2017–N–6565] RIN 0910–AH60. *Fed Regist.* 2018;83(55): 12294–12301.
128. Department of Health and Human Services Food and Drug Administration. Menthol in cigarettes, tobacco products; request for comments 21 CFR Part 1140 [Docket No. FDA–2013–N–0521]. *Fed Regist.* 2013;78(142):44484–44485.
129. Department of Health and Human Services Food and Drug Administration. Tobacco product standard for nicotine level of combusted cigarettes 21 CFR Part 1130 [Docket No. FDA–2017–N–6189] RIN 0910–AH86. *Fed Regist.* 2018;83(52): 11818–11843.
130. Chaloupka FJ, Cummings KM, Morley C, Horan J. Tax, price and cigarette smoking: evidence from the tobacco documents and implications for tobacco company marketing strategies. *Tob Control.* 2002;11:62–72. https://doi.org/10.1136/tc.11.suppl_1.i62.
131. Campaign for Tobacco-Free Kids. Tobacco Product Marketing on the Internet. Washington DC https://www.google.com/url?sa=t&rct=j&q=&esrc=s&source=web&cd=1&cad=rja&uact=8&ved=2ahUKEwigo5q2kYXjAhUk1VkKHfeyCEwQFjAAegQIARAC&url=https%3A%2F%2Fwww.tobaccofreekids.org%2Fassets%2Ffactsheets%2F0081.pdf&usg=AOvVaw2lnit2F8c31_8fH0xsf6Ag.
132. U.S. Food and Drug Administration. *The Public Health Rationale for Recommended Restrictions on New Tobacco Product Labeling, Advertising, Marketing, and Promotion*; 2019. Washington, D.C https://www.fda.gov/media/124174/download.
133. Kaur J, Jain DC. Tobacco control policies in India: implementation and challenges. *Indian J Public Health.* 2011;55(3):220–227. https://doi.org/10.4103/0019-557X.89941.
134. McKay AJ, Patel RKK, Majeed A. Strategies for tobacco control in India: a systematic review. *PLoS One.* 2015;10(4):e0122610. https://doi.org/10.1371/journal.pone.0122610.
135. Gartner C, Hall W. Harm reduction policies for tobacco users. *Int J Drug Policy.* 2010; 21(2):129–130. https://doi.org/10.1016/J.DRUGPO.2009.10.008.
136. Zeller M, Hatsukami D. Strategic Dialogue on Tobacco Harm Reduction Group. The strategic dialogue on tobacco harm reduction: a vision and blueprint for action in the US. *Tob Control.* 2009;18(4):324–332. https://doi.org/10.1136/tc.2008.027318.
137. Levy DT, Mays D, Boyle RG, Tam J, Chaloupka FJ. The effect of tobacco control policies on US smokeless tobacco use: a structured review. *Nicotine Tob Res.* 2017; 20(1):3–11. https://doi.org/10.1093/ntr/ntw291.

CHAPTER THREE

Epidemiology of smokeless tobacco use in the United States and other countries

Gretchen McHenry, Stephanie J. Parker, Dalia Khoury, Kevin P. Conway
RTI International, Research Triangle Park, NC, United States

Introduction and definition

Smokeless tobacco (ST) is widely used in the United States and internationally. The World Health Organization (WHO) defines ST as "tobacco products without combustion or pyrolysis at the time of use."[1] This definition does not include nicotine or tobacco products that are consumed via electronic delivery systems ("vaped").

ST takes many forms. They include snuff, chewing tobacco, snus, dip, and any oral or nasal tobacco combined with other substances, such as areca nut or betel quid. Because of its many forms and variations in rates of use across populations, definitions of ST vary across epidemiologic studies. Table 3.1 points out general definitional differences among several large studies that will be examined more closely in this chapter. There are considerably more variations of ST internationally than domestically, particularly in South-East Asia where ST is often mixed with other substances such as betel quid or areca nut. (In this chapter, South-East Asia follows the WHO definition: Bangladesh, Bhutan, Democratic People's Republic of Korea, India, Indonesia, Maldives, Myanmar, Nepal, Sri Lanka, Thailand, and Timor-Leste).

This chapter considers a broad range of ST products. It is important to keep in mind that reports of prevalence and trends across surveys may differ because of definitional differences. Sample sizes, age groups, and time period of use may also vary by survey. These differences will be considered throughout. The literature addressing characteristics of use, such as polyuse of tobacco products, as well as the risk factors, correlates of use, and health outcomes linked to ST use will also be covered.

Smokeless Tobacco Products
ISBN: 978-0-12-818158-4
https://doi.org/10.1016/B978-0-12-818158-4.00003-0

Common selected forms or smokeless tobacco: Domestic and international

Selected Forms of Smokeless Tobacco Domestic and International	Description
Chewing tobacco	Tobacco that is chewed or held in the mouth, typically between the cheek and gums, rather than smoked. Can be loose leaf, plug, twist. Can also be referred to as dipping.
Snuff	Finely cut or powdered flavored tobacco. Can be prepared as moist or dry snuff or packaged in sachets.
Snus	Finely ground moist snuff that can be loose or packaged.
Dissolvable tobacco/ "dry snuff"	Tobacco that dissolves in the mouth. Can be distributed sticks (twisted sticks in the size of a toothpick), strips (film strips for the tongue), or orbs (pellets).
Paan	Mixture of betel quid with areca nut, with or without tobacco. May also be mixed with slaked lime, katha paste, or mukhwas.
Gul/gutkha	Oral tobacco powder that is spread over the gums and teeth. Can also contain crushed areca nut, catechu, paraffin wax, slaked lime, and/or other sweet or savory flavoring.
Toombak	Finely ground tobacco leaves mixed with "atron" (sodium bicarbonate) and water.
Naswar	Moist, powered tobacco similar to snus.

Literature collection

Works included in this review were found via a systematic process which included searching electronic databases and websites of organizations involved in tobacco research. Specific surveys that measure ST use and its risk factors, correlates, and health outcomes were also examined. Primary data analysis is included for recently released data sources which do not yet have a strong presence in the literature.

Electronic databases, including PubMed, PsycINFO, and Web of Science, were searched for relevant articles. The search was restricted to articles from 1998 to 2018 in order to present the most recent information. Citations from the articles found in these databases were also considered for

Table 3.1 Definitions of smokeless tobacco by survey.

Major epidemiological studies	Definition
United States	
Monitoring the Future (MTF)	Combines snuff and chewing tobacco
National Survey on Drug Use and Health (NSDUH)	Combines snuff and chewing tobacco
Youth Risk Behavior Survey (YRBS)	Combines snuff, chewing tobacco, dip, snus, and dissolvable tobacco
Behavior Risk Factor Surveillance System (BRFSS)	Combines snuff, chewing tobacco, dip, and snus
National Health and Nutrition Examination Survey (NHANES)	Measures snuff and chewing tobacco separately
Population Assessment of Tobacco and Health (PATH)	Combines chewing tobacco, snuff, dip, and snus
International	
Global Adult Tobacco Survey (GATS)	Combines snuff, chewing tobacco, and snus
Special Eurobarometer	Combines snuff and chewing tobacco

inclusion. A reverse citation search was conducted on key articles, and the relevant works found were also included.

Domestic prevalence and trends

The prevalence estimates of ST use in the United States have ebbed and flowed. It was the most popular form of tobacco throughout the early 20th century. As cigarettes became more popular and reached a peak in the 1960s, ST use declined[2] and was low from the late 1980s to the early 2000s. It did not decline across all subgroups, however, during this period; use increased among young white men, particularly when tobacco marketing focused on ST in that demographic.[3,4] Also, Maher et al. found that use remained stable among Alaskan men between 1996 and 2008; in fact, it increased for those who were current smokers.[5]

Recent national level data show a prevalence for current and recent use ranging from 1.3% to 9.1%, depending on survey year, definition, and age group. Generally, estimates have been stable or show small changes (some increases, some decreases) across certain years. The National Adult Tobacco

Survey (NATS), a nationally representative telephone survey with adults aged 18 and older, measured current use of ST as now using chewing tobacco or snus every day or on some days. In 2012, NATS yielded a current prevalence rate of 1.7%.[6] Expanding beyond just adults, Wave 1 of the Population Assessment of Tobacco and Health (PATH) Study, a nationwide longitudinal household survey collected in 2013 and 2014 with respondents aged 12 and older, defined current users as those who use either every day or some days. Prevalence for current use was 3.4% for adults and 1.6% for youths.[7] Also for 2014, the National Survey on Drug Use and Health (NSDUH), a nationally representative household survey conducted with respondents aged 12 and older, reported a similar past-30-day prevalence for youths, 2.0%. For adults in the United States, a 2.3% past-30-day prevalence was reported in 2014 in the Tobacco Products Risk Perception Surveys (TPRPS). The TPRPS is an annual web survey conducted with noninstitutionalized adults (aged 18 and older) living in the United States. The TPRPS also reported a significant increase in past-30-day prevalence in 2015, rising to 3.7%, then a significant decrease in 2016, dropping back to 2.7%.[8] For all respondents aged 12 and older, NSDUH reported a 3.3% past-30-day prevalence in 2014.[9]

To better understand ST use and where opportunities for change may exist, it is important to examine use among different geographical regions and key subgroups, and how use is associated with risk factors and health outcomes.

International prevalence and trends

Use in countries outside the United States varies broadly. This variance can be seen in overall prevalence, prevalence by the type of ST, and use trends. Some of these differences are cultural and reflect the tobacco and other substances most readily available in different parts of the world. Other differences may be driven by national policy on tobacco use. For example, the United States does not advocate use of ST as a smoking cessation method, while many other countries do. In addition, the belief that ST is a potentially less harmful alternative to combustible tobacco[10] likely increases prevalence and impacts use trends.

Europe

Overall, ST use is low throughout Europe, relative to the United States and Asia. Two recent studies, the 2010 Pricing Policy and Control of Tobacco in

Europe (PPACTE) project and the 2012 Special Eurobarometer 385, reported an average of 1.7% and 2%, respectively, of current use for respondents aged 15 and older across European countries.[11] A notable exception to this low prevalence is Sweden.

Snus is the most commonly available form in Sweden. The manufacturing process makes it lower in nitrosamines, a carcinogen specific to tobacco, than in other forms of ST.[12] Its use has been increasing over the past 25 years, while use rates for other forms of tobacco have been falling. This is likely due to Sweden's acceptance of snus as a smoking alternative.[13] A potentially confounding factor is that snus has been illegal to sell in the European Union (EU) since 1992, with the exception of Sweden. Possession or use of snus is not illegal in other European countries, such as Norway, but the inability to easily purchase it likely drives down use.[14]

Reports of current ST use in Sweden vary across surveys, but the estimates are higher than in the rest of Europe. The 2010 PPACTE, a household survey conducted in 18 European countries with respondents aged 15 and older, reported that 12.3% of respondents in Sweden were regular users.[11] Agaku et al.[15] found a higher rate by combining data from the 2008–12 Global Adult Tobacco Survey (GATS) and the 2012 Special Eurobarometer 385. Both are nationally representative household surveys conducted with respondents aged 15 and older. The GATS was done in more than 25 low- and middle-income countries while the Eurobarometer 385 was done in 27 EU countries. The GATS had a broad definition of current use. Respondents who indicated daily or less than daily use in response to the question "Do you currently use smokeless tobacco on a daily basis, less than daily, or not at all?" were considered current users. The Eurobarometer 385 considered current users to be anyone who indicated regular or occasional use. This analysis showed a 17.7% current use rate for Swedish respondents. This broad definition could contribute to the higher prevalence than that seen in the 2010 PPACTE, which explicitly asked if a respondent was a current user.

Asia

ST use is highly prevalent across Asia. In a WHO study of 121 countries, 10 in South-East Asia accounted for 82.7% of total use.[16] Many Asian countries have a wider variety of ST products available than western countries. Popular forms include paan, paan masala, zarda, gutkha, betel quid with tobacco, and khaini.[17,18] There has been a decrease in production of ST in Asia over the last 70 years, but few countries show a decline in use.[17] Despite this decrease in production, use is rising in many countries, including India, Bangladesh, and Myanmar.[17–20] There is very little regulation of ST production in most Asian countries, making control programs difficult.[20]

In India, although the prevalence of ST use varies widely by geographic area, overall 25.9% of Indians aged 15 and older are current users according to the 2009 GATS.[17,18] The most popular form is khaini, which is a tobacco and lime mix; it is consumed by 11.6% of current users.[17] Other common forms include betel quid, gutkha, and paan masala.[21] Many of the common forms are shown to have higher toxicity than products in western markets.[22] Kostoya and Dave[23] utilized the 2009 GATS data to measure the intensity of any ST use. They found that current users averaged 6.6 times per day. Considering the elevated prevalence in India, along with the high level of toxicity and intensity of use, ST likely poses a significant public health risk.

Use in Bangladesh is higher than in India. According to the 2009 GATS, 27.2% of Bangladeshis are current users.[15] This is the highest reported rate in Asia and the world, and it is increasing. In 2004, the rate for adults was 19.7%; it increased by 7.5% in just 5 years. A variety of products are available. Popular types are betel quid, paan, paan masala, zarda, and gutkha.[18] Bangladesh faces the same issues as India when it comes to control.[20]

South-East Asian countries where smokeless tobacco is highly prevalent

South-East Asian countries that account for 82.7% of total smokeless tobacco use	
Bangladesh	Myanmar
Bhutan	Nepal
India	Sri Lanka
Indonesia	Thailand
Maldives	Timor-Leste

Domestic and international risk factors

Poly-use of tobacco

The magnitude of poly-use of tobacco varies by its operational definition.[24] Thresholds for poly-use have been set at any use of more than one product in the past year, or past month, or frequency according to respondent reports. Poly-use may be defined as concurrent use, meaning current use of more than one tobacco product. Some studies also consider intensity of use. For example, a poly-user may be defined as a person who uses more than one tobacco product every day or some days. Different definitions can greatly affect prevalence rates and analyses.

Analyses often combine all noncigarette products or even all nonmanufactured cigarettes (such as bidis, kreteks, or any form of roll-your-own) into one category. A weakness of this approach is that it suggests that noncigarette tobacco use is relatively uncommon, which is not the case in many countries. For example, both Sweden and India have higher rates for ST than for manufactured cigarettes.[15] According to Wave 1 of the PATH Study, conducted in the United States with adult respondents, cigarettes are still the most common form of tobacco, with a current prevalence of about 18%.[25] Because of this tendency to consider poly-use as use of cigarettes and any noncigarette product, it is difficult to isolate poly-use involving ST in the literature. Throughout this section, poly-use is treated as a general topic, and studies specifically focused on ST are highlighted.

United States

The concept of poly-use as a risk factor is relatively recent. Bombard et al.[26] first used 2003 Behavior Risk Factor Surveillance System (BRFSS) data to examine it. By looking at data from 10 states, they found that 3.4% of adults aged 18 and older were poly-users of cigarettes and at least one other tobacco product. Though Bombard et al.[26] were the first to examine poly-use by comparing cigarettes to a combination of other tobacco products, a previous study estimated that concurrent use of cigarettes and ST was low, at less than 1.0%.[27] Some special populations had a higher prevalence, such as recently enlisted Air Force members. They reported a 13.8% rate of current dual-use of ST and cigarettes.[28] Dual-use of ST and cigarettes is a common combination. In 2011, the Current Population Survey Tobacco Use Supplement (TUS-CPS) found that 2.8% of current adult cigarette smokers were also every day or some day users of ST.[29] Among all American adults, the dual-use of ST, defined as chewing tobacco and snus, and cigarettes in 2012 was 0.6% in the NATS.[6] There is evidence that dual-use has been increasing over time, corresponding with an increase in ST advertising and sales.[30]

Though dual-use of cigarettes and ST is not highly prevalent in the general population, there are some key factors which affect dual-use prevalence. The 2009 BRFSS showed that current adult smokers who also use ST range from 0.9% (in Puerto Rico) to 13.7% (in Wyoming). This type of poly-use was highest among 18- to 24-year-olds.[31] Among male daily ST users over 25, 7.3% also smoked daily.[32] For a broader comparison of adults, the ConsumerStyles survey, a nationally representative mail survey conducted with respondents aged 18 and older, showed that 41.3% of daily or some day ST users also smoked daily or on some days.[31] According to PATH data from 2013 to 2014, 4.0% of adults who report current use of two or more tobacco products currently used cigarettes, defined as smoking more than 100 cigarettes in their lifetime and currently smoking every day or some days, and used ST products other than snus. An additional 1.0% of adult respondents indicating poly-use reported currently traditional cigars and ST products other than snus. Another 1.0% of adult poly-users reported using ST and snus.[7]

Adolescents have a much higher prevalence of poly-use. According to the 2004 National Youth Tobacco Survey (NYTS), 69.1% of middle school boys who used ST daily also smoked in the past month, with 53.8% smoking daily. That increased prevalence of dual-use held for high school aged males who used ST daily. Sixty percent of high school boys who used ST in the past month also smoked cigarettes in the past month. Intensity of dual-use increased with age throughout adolescence.[32] Osibogun et al.[33] looked more closely at youths and young adult poly-users. They found that 3.6% of 12- to 17-year-olds, 21.7% of 18- to 24-year-olds, and 15.8% of 24- to 34-year-olds were poly-users of at least two forms of tobacco in the past month. Respondents aged 12 to 17 in the PATH Study reported using ST products in a variety of combinations with other tobacco types. Current use of cigarettes and ST products other than snus was reported by 4.0% of adolescent poly-users. Three percent of adolescent poly-users reported using ST and snus. Another 3.0% of adolescent poly-users reported current use of cigarettes, e-cigarettes, and ST products other than snus.[7] This higher prevalence of use of e-cigarettes combined with ST is unique to adolescent poly-users.

Overall, for both adults and adolescents, poly-users are more likely to be male, white, and non-Hispanic, live in rural areas; live in the South or Midwest; and have lower educational attainment and lower income.[26,31,32,34,35] Poly-use is also correlated with heavy drinking and high-risk behaviors.[26,34,36] There is some evidence that heavy drinking increases the odds of poly-use in women slightly more than it does in men.[34] Though men and younger adults are more likely to be poly-users, Mushtaq et al.[34] found that the percent of dual-users of ST and cigarettes evens out by gender in the 35- to 44-year-old age group, and that a higher percentage of women over 45 are dual-users than men. There is little research on women ST users and poly-users largely because they are less prevalent in the population; this finding points to a need for more research on women and other high-risk groups that may be overlooked because they are less common.

Attitudes of poly-users about tobacco use differ from those who only smoke or only use one other form. Dual-users of ST and cigarettes are less likely to report planning to quit using tobacco than those who only smoke. Roughly half of dual-users do not plan to quit, and 75.1% believe using ST would not help them quit smoking.[31] Poly-use may be driven by a perception of lower harm associated with tobacco products other than cigarettes, which those who do not poly-use may not share.[33]

International

Poly-use varies widely by country and region. In Denmark, 11.9% of the population are poly-users. India has the highest prevalence of current smokers who also use at least one other tobacco product, at 66.2%. Worldwide, 28 countries have at least 20% of their smokers also using at least one other tobacco product.[15] It is important to note that many international studies distinguish between manufactured cigarettes and hand-rolled cigarettes because the latter are more common in countries other than the United States.

Across 44 countries located in all 6 WHO regions, adults over the age of 25 are more likely to be poly-users than those under 25.[15] This is a key difference from the United States. However, similarly to the United States, men and those living in rural areas are more likely to be poly-users. Those from South-East Asia were more likely to be poly-users than those in Europe. However, those from other parts of the world, including the Americas, were less likely to be poly-users than those in Europe.[15]

Detailed studies of poly-use and dual-use of ST and another tobacco product are not common. Lund and McNeil[10] examined dual-use of cigarettes and snus among men aged 16 to 74 in Norway. This focus is especially important because snus sales make up 30% of tobacco sales in Norway. Lund and McNeil[10] defined dual-use as daily use of one product and at least weekly use of the other product. They found that 6.8% of Norwegian men were dual-users, with 1.0% using both products daily. Of daily snus users, 9.8% also smoke cigarettes daily. Approximately 41% of snus users who reported "occasional" use also smoke daily. The prevalence of dual-use in Norway is similar to that reported in Sweden, which has a similar snus market.[10]

Dual-users aged 16 to 74 in Norway smoked at a lower intensity, roughly three cigarettes per week fewer, than cigarette users who did not use ST. Also, an increase in snus use has not led to an increase in dual-use of cigarettes and snus. Though advertising snus is illegal, and the Norwegian government does not present snus as a safer alternative to smoking, respondents most often cited smoking reduction and substitution as their reasons for dual-use. These reasons were significantly more likely to be reported than smoking cessation, which suggests that Norwegian men may be attempting to mitigate risk. Dual-use did not negatively affect respondents' plans to quit smoking.[10]

Those who use more than one tobacco product may be at greater risk of negative health outcomes than single product users. These outcomes may differ in magnitude or in form from single product users.[15] There are unique health risks associated with the dual-use of cigarettes and ST. Teo et al.[37] found an increased risk of myocardial infarction (MI) in dual-users when compared to users of either product individually. Without more studies that separate noncigarette tobacco products into more specific analytic categories, information about those potential health outcomes will remain inadequate.

Flavored smokeless tobacco products

In the United States, flavored tobacco products saw an increase of 65.6% in sales between 2005 and 2011.[38,39] Flavored moist snuff, including snus, increased 72.1% in the same period, representing 60% of the growth in flavored products overall.[38] This increase in sales of flavored products is concerning because previous research shows that flavoring attracts new users and youths to tobacco products.[38,40]

In the United States, 70% of middle and high school students who are current tobacco users use at least one flavored tobacco product.[41] Over

69% of youths aged 12 to 17 who use ST report that they do it because of added flavoring. This was the second highest reported reason, immediately following the ability to use ST where smoking is prohibited.[42] Mint flavoring was the most popular among youths (aged 12 to 17) and young adults (aged 18 to 24).[38,43] In a meta-analysis of five previous studies conducted with adult ST users, Oliver et al.[43] found that the majority of adult users reported that their first ST product, first regularly used ST product, first daily ST product, and current ST product were all flavored. This analysis also found that 51.3% of users whose first ST product was not flavored switched to a flavored product by the time of the study.[43]

Flavoring could contribute to ST initiation in youths because it is perceived to be easier to use, as indicated by 20.2% of PATH Study respondents aged 12 to 17. Another 44.3% believed flavored ST products had the same ease of use as unflavored products.[44] Analysis of PATH data also showed that 81% of respondents aged 12 to 17 who had ever used ST first used flavored ST. Of ever ST users who initiated use before age 15, 80% first used flavored ST products.[40]

More research is needed on flavored tobacco products, including ST. In the United States, the Food and Drug Administration (FDA) and National Institutes of Health (NIH) are currently funding Tobacco Centers of Regulatory Science (TCORS) grants. As work on these grants continues, more research that includes issues related to flavored ST use may be published.

Internationally, there is a lack of published research, which may be due to the wide range of products and how they are manufactured in other countries, particularly in South-East Asia. However, in Canada, there has been a growing trend to advertise and develop flavored products aimed at youths. This is believed to have led to increased experimentation with flavored tobacco products by youths and a call to ban them throughout Canada, as seen in the 2004–08 Youth Smoking Survey (YSS).[45]

Demographic risk groups and special populations

Those who use ST either exclusively or in conjunction with other tobacco products are at increased risk of negative health outcomes as compared to nonusers. ST use is not evenly distributed throughout the population, either in the United States or internationally. Characteristics such as sex, age, race and ethnicity, income level, and urbanicity affect prevalence and need to be considered when discussing health outcomes and public policy.

United States

Men are significantly more likely than women to use ST.[3,8,46,47] In 2002, Howard-Pitney and Winkleby[3] reported that men aged 25 to 64 used ST products at a rate 10 times higher than women in the same age group. That gap appears to have grown. PATH Wave 1 data show that 5.7% of adult men currently use ST while only 0.2% of adult women do.[47] Jones et al.[8] found slightly different prevalence estimates for male and female current users in data from the 2014–16 Tobacco Products and Risk Perceptions Surveys (TPRPS). In 2014, 4.1% of adult men and 0.6% of adult women currently used ST products. In 2016, those numbers slightly decreased to 3.8% for men and increased to 1.7% for women.[8] Part of the discrepancy between these two studies may be definitional. Current use is defined as using some or every day in PATH but as any use in the past 30 days in the TPRPS.[8,47] Jones et al.[8] also suggest that women's use may be rising due to an increased marketing push toward them. Examining details on women's use is difficult because most studies restrict themselves to deeper inspection of use among men only.

Fig. 3.1 illustrates this striking gender gap in the NSDUH. Rates are not only orders-of-magnitude higher among males than females regardless of age group, but the higher rates for males are consistent at 10-year intervals across 20 years. Younger adults are more likely to use than older

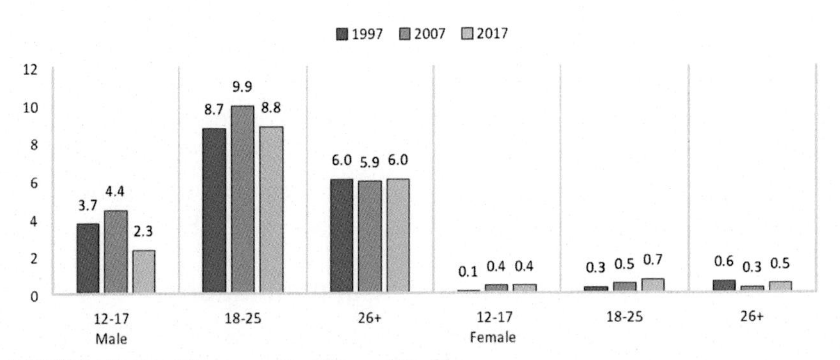

NSDUH = National Survey on Drug Use and Health
Note: Data for this figure from NSDUH Detailed Tables publications and archived ICPRS NSDUH datasets.

Figure 3.1 NSDUH smokeless tobacco past month prevalence of use by age and gender.

adults[7,47,48] or youths. This is evident in Fig. 3.1, particularly for men, in the spike in use for respondents aged 18 to 25, relative to the other age groups.

Age is a major driver of ST use across studies. In the 2009 BRFSS, those aged 18 to 24 had the highest prevalence of current use.[48] Other studies have shown similar results with slightly different age ranges. The Wave 1 of PATH data showed those 18 to 24 and 25 to 34 each having a prevalence rate of 4.0%; that number dropped to just 1.6% for those aged 50 and over.[47] In the TPRPS data, those 18 to 29 had a current prevalence of 2.6%; the rate fell evenly across age groups until taking a bigger dive down to 1.3% for those 60 and over. The age group most at risk and the pattern of use across age groups remain consistent across varying definitions of current use and different ranges of age categories. Fig. 3.2 uses NSDUH data to give an example of how ST use has changed over time for three age groups: adolescents (aged 12 to 17), young adults (aged 18 to 25), and adults (aged 26 and older). This figure clearly shows that, for the past three decades in the United States, ST use is consistently and substantially more prevalent among young adults than other age groups.

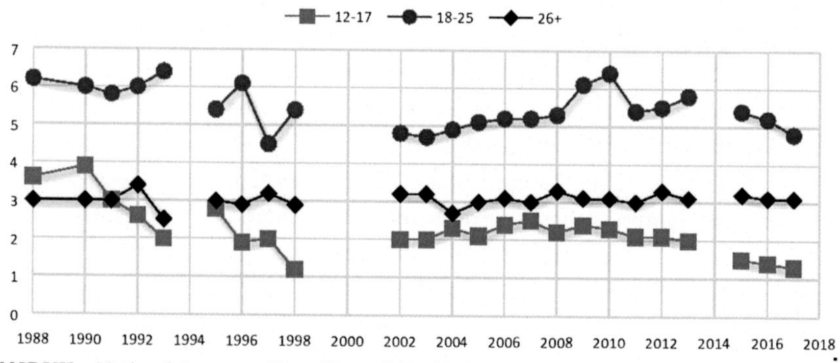

NSDUH = National Survey on Drug Use and Health

Note: NSDUH data for youths aged 12 to 17 are not presented for 1994, 1999 to 2001, and 2014 due to design changes in the survey, which preclude direct comparisons of estimates with subsequent years.

Note: Data for this figure from NSDUH Detailed Tables publications and archived ICPRS NSDUH datasets.

Figure 3.2 NSDUH smokeless tobacco past month prevalence by age group and survey year: 1988–2017.

Despite higher estimates of use among young adults (aged 18 to 25), adolescence is a period of elevated risk of ST use. Adolescence is a time of great change and is often when substance use, including ST, begins. According to the Adolescent Health Risk Behavior Study (AHRBS) 9% of middle and high school students in one county in Indiana had ever used ST.[49] In 2017, Monitoring the Future (MTF), a large nationally representative survey conducted in schools with 8th, 10th, and 12th graders, found that past-30-day prevalence increased with age: 1.7% for 8th graders, 3.4% for 10th graders, and 4.8% for 12th graders.[50] Similar to MTF's 8th graders, the NYTS found a past-30-day prevalence for all middle schoolers of 1.9%. However, NYTS found a higher past-30-day prevalence of 5.5% for all high school students.[51] This is likely due to sample differences between MTF and NYTS. Holman et al.[52] found in a longitudinal study that 9.4% of male respondents who had never used at age 12 were daily users at 18. Of those, 20% had quit by age 20.[52] Adolescents are at an increased risk due to a range of factors, including poor impulse control, perceived peer approval, and peer influence.[49,52]

Rates of use among American youths have largely decreased over the past three decades. Fig. 3.3 shows past month use among youths for three major national studies conducted annually from 1992 to 2017. Despite some variations over time among certain age groups, rates of use decreased markedly from 1992 to 2002. The decreased level of ST use achieved in the early 2000s has remained generally stable for more than a decade, with some exceptions. Signs of continued decreasing are appearing in recent years.[46] Overall, the rates of ST use among American youths in each age group decreased by at least 50% from 1992 to 2017.

Most ST users in the United States are non-Hispanic and white.[3,46,47] There is some evidence that American Indian and Alaskan Natives have current rates that match or exceed those of whites, but that information is not widely reported.[3] Often, race is condensed to two categories for analysis, white and nonwhite, because of small sample size. This may mask a growing prevalence of use in individual racial groups other than whites. Jones et al.[8] looked at four categories: white non-Hispanic, black non-Hispanic, other non-Hispanic, and Hispanic. They found that black non-Hispanic respondents had the highest rate of current use at 3.5%. White non-Hispanic rate was 2.3%, which is lower than seen in many other studies, while other non-Hispanic and Hispanic rates were 2.3% and 2.2%, respectively.[8] This is a unique finding and calls for future research to consider race and ethnicity in more varied categories.

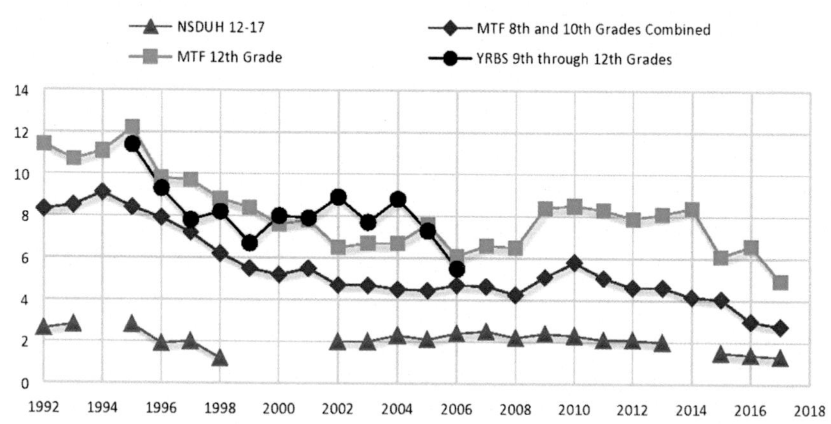

MTF = Monitoring the Future; NSDUH = National Survey on Drug Use and Health; YRBS = Youth Risk Behavior Survey.

Note: For comparison purposes, MTF data on 8th and 10th graders are combined to give an age range close to 12 to 17 years, the standard youth age group for NSDUH.

Note: NSDUH data for youths aged 12 to 17 are not presented for 1994, 1999 to 2001, and 2014 due to design changes in the survey, which preclude direct comparisons of estimates with subsequent years.

Note: Data for this figure comes from *Monitoring the Future national survey results on drug use: 1975-2017: Overview, key findings on adolescent drug use, 2017 Youth Risk Behavior Surveillance Summary, Trends in the Prevalence of Tobacco Use National YRBS: 1991—2015* (2016), the 2014 and 2017 *National Survey on Drug Use and Health: Detailed Tables*, and ICPSR archived NSDUH datasets.

Figure 3.3 Past month smokeless tobacco use among youths in the United States from 1992 to 2017.

Exclusive ST use does not appear to have a significant relationship to income but does have a correlation with other measures of social class.[8,53] For example, men with lower levels of education are more likely to use than those with higher levels.[46,47] Another factor that has a positive correlation with use is blue collar employment.[53,54] Blue collar workers had a higher than average current use rate at 9.5%. This was even higher in the construction trades, with prevalence of 35% exclusive ST and 37% dual-use with cigarettes.[54] Career firefighters and military members also had a higher rate than the general population. Firefighters have an exclusive ST rate of 13.3% and a lower prevalence of dual-use with cigarettes at 2.6%.[55] Recently enlisted Air Force members use ST at roughly twice the rate of those not in the Air Force; dual-use of cigarette and ST among recruits was 13.8%.[28]

Users are more often rural than urban dwellers.[46,47,56] Over 8% of adult nonurban residents were current users in 2013–14, compared to only 2.5% of adult urban residents.[47] Use in rural adolescents may be even higher. Couch et al.[56] found that 32.7% of adolescent males in rural California were current users, and 7.2% considered themselves former users. Less than half had never tried ST.[56] The higher rates for rural residents, particularly men, may be driven by cultural norms and views of masculinity.[57] Though this population currently has relatively high rates, they have dropped since the 1990s. In 1992 and 1993, 9.8% of adult men living outside a metropolitan statistical area used ST. That number fell to 5.9% in 2010 and 2011.[46] The variation is likely due to definitional differences in what makes up a nonurban area and on the population studied. Nevertheless, geography is a major factor in ST rates in the United States. Fig. 3.4 shows current use by state in 2016.

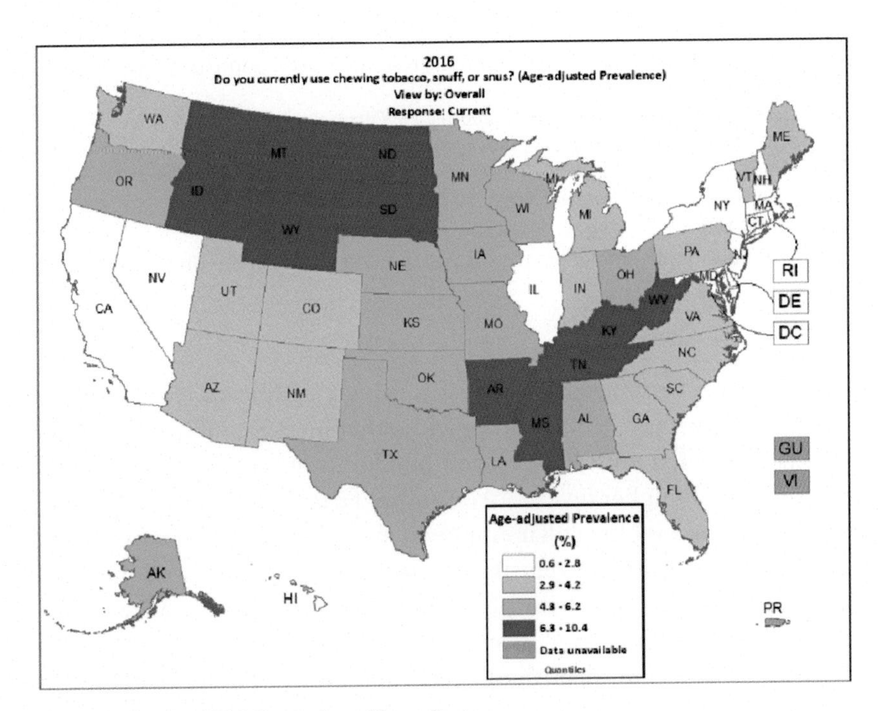

BRFSS = Behavioral Risk Factor Surveillance System
Note: Prevalence estimates are not available if the unweighted sample size for the denominator was < 50 or the Relative Standard Error (RSE) is > 0.3 or the state did not collect data for that calendar year.
Note: Data for this figure from the *2016 BRFSS Prevalence & Trends Data* publication.

Figure 3.4 Current smokeless tobacco use by state in 2016 (BRFSS).

A positive association has been reported between tobacco use and both substance use and mental health problems.[58] The focus of these studies is principally on cigarettes, but a positive association has also been shown for adult ST users. Adults who had ever used ST, defined as chewing tobacco and snuff, had a higher prevalence of mood disorders than those who had not. The only significant increase was in instances of mania or hypomania, but rates of major depression and dysthymia were also higher for any lifetime ST users compared to nonusers. Users also had higher rates of personality disorders and substance use disorders.

The strength of these associations varied dependent on the type of ST product.[59] Those who used in the past year, as opposed to any lifetime use, also had a positive association with mental health and substance use

Personality and substance use disorders associated with smokeless tobacco use

Personality disorders:
- Avoidant
- Dependent
- Obsessive–Compulsive*
- Paranoid*
- Schizophrenic*
- Antisocial*
- Histrionic*

Substance user disorders:
- Alcohol*
- Nicotine*
- Cannabis*
- Amphetamine*
- Opiate*
- Sedative*
- Tranquilizer*
- Inhalant/solvent*
- Cocaine*
- Hallucinogen*
- Heroin*

*Statistically significant relationship.

disorders. Past year users were twice as likely as nonusers to have a high severity substance use problem. They were also more likely to report both internalizing and externalizing mental health problems.[58] The association of ST use with substance use disorders and mental health issues may have differential impacts on certain populations. For example, military members who were deployed in Iraq or Afghanistan had a higher rate of ST initiation and persistent use. Those with PTSD symptoms were more likely than those without to use. The increased usage by deployed troops, particularly those with PTSD symptoms, may be a coping mechanism for increased stress.[60] Similarly, Native Americans were 1.6 times more likely to use if they also showed PTSD symptoms. Lifetime use within native populations is high, but rates increased for those who also had psychological disorders.[61]

International

The demographics of use and risk factors for use in other parts of the world differ from those in the United States. Though most users worldwide are male, the rate of female use is higher in some countries than in the United States. In Sweden, 3.1% of adult women and 3.5% of adolescent girls were daily or occasional snus users,[62,63] while the rate of daily use among men was 17.0% for the same period.[62] However, Agaku et al.[15] found a higher prevalence: current use of any ST was 27.7% for men and 7.9% for women. The pattern of higher use among men than women is reversed in several places in South-East Asia. India has the familiar pattern, 32.9% of men and 18.4% of women. But in Thailand, Bangladesh, and Indonesia, the rate is higher in women[15] (Table 3.2).

Similar to the United States, younger adults (under 35) are more likely to use snus than older adults in Sweden.[62] Recent years have seen an increase in use among adolescents.[63,64] In 2008, 21.5% of boys and 3.5% of girls were daily or occasional users of snus. Adolescents also had high rates of dual-use with cigarettes compared to adults.[63] Unlike the United States and Sweden, ST users in many parts of South-East Asia are more likely to be older. In

Table 3.2 Current use of smokeless tobacco by gender in three South-East Asian countries.

Country	Men (%)	Women (%)
Thailand	1.3	6.3
Bangladesh	26.4	27.9
Indonesia	1.6	2.0

Sample includes respondents aged 15 and older.

Bangladesh, India, and Thailand, older adults are more likely to use any form of ST than younger adults. In each country, the age group with the highest prevalence was 65 and over.[17]

For men in Sweden, snus use is associated with several socioeconomic factors. Skilled workers, like tradesmen, are more likely to be daily users than those in other professions, as were men with low and intermediate education and with intermediate income. Snus use among women did not have an association with income or occupation, but women with intermediate education were more likely to be daily users than others.[62] These associations are only true of adults; there was no association between socioeconomic factors and adolescent use.[64] In India, Bangladesh, and Thailand, a higher proportion of lower income and lower educated individuals are current ST users, and rural residents are more likely to use than those who live in urban areas.[17]

Health outcomes

The presence and magnitude of negative health outcomes associated with ST use are greatly debated. There is evidence that it increases the likelihood of dental disease, various cancers, cardiovascular disease, negative pregnancy outcomes, and overall mortality.[20,65–67] However, analyses of health outcomes in the United States are limited by the low prevalence of ST use in the general population. Longitudinal studies, particularly in high-use populations, could give a richer picture,[65] but even they suffer because of changes in product consistency over time.[67]

Though use is higher in Asia, products there are known to have higher toxicity and rates of cancer-causing chemicals, making results incomparable to those in other parts of the world.[65] Because of the greater prevalence and

Countries with elevated rates of smokeless tobacco use among adults over age 65

	Smokeless tobacco use for those over 65 years old (%)
Bangladesh	56.4
India	33.7
Thailand	13.6

toxicity, an estimated 85%–88% of the global disease burden occurs in South-East Asia.[20,66] Siddiqi et al.[66] found that, in 2010, 1.7 million disability-adjusted life years (DALYs) were lost and over 62,000 deaths could be attributed to cancers associated with ST use globally. The rates for heart disease were even higher: 4.7 million DALYs and over 200,000 lives. Sinha et al.[20] calculated that 9% of all deaths and 23% of all DALYs lost associated with any tobacco use were affected by ST use. As global use increases, the negative health outcomes will increase and will have differential effects in various parts of the world.

There is also an argument that ST use should be encouraged as a means to stop or reduce smoking.[10,68–70] Relative to cigarette use, the negative health outcomes associated with ST are less severe.[68–70] There is a lower risk of cancer than with cigarettes, but some evidence that ST can reduce the cancer survival rate. There is also limited evidence that it may increase the risk of fatal acute MI compared to cigarettes.[70] Some countries may benefit form promoting ST as a tool to aid in smoking cessation, but this approach is not likely to lower the disease burden in countries with ST products that have high toxicity or where prevalence is already high.

Dental disease

ST use has been linked to an increase in dental caries in the United States and Sweden.[65] There is also a positive association with periodontal disease. Users are twice as likely as those who have never used to develop severe active periodontal disease and interproximal attachment loss. Cigarette smoking also is associated with these disorders, but the link to ST remained after controlling for cigarette use.[71] A study of men in a rural area in the United States who use ST daily found greater gingival recession in prevalence and extent at the sites where ST was placed inside the mouth. These findings were only significant for those who had used for more than 10 years; those who had used for fewer than 10 years did not differ significantly from nonusers. Daily users also had more plaque and gingivitis despite 88% of the sample reporting daily brushing. Greater attachment loss, a sign of irreversible periodontal disease, was observed on teeth at the placement site for daily users.[72] Though there is consistent research showing increased odds of dental disease, the sample size for many of these studies was small and calls into the question the generalizability of their conclusions.[65]

Cancer

The risk of cancer varies widely with the type of product, which, as previously described, correlates strongly to region. Research has focused on cancers of the head and neck (oral, pharyngeal, and esophageal) and the pancreas. A handful of studies have found weak positive or nonsignificant association with other forms of cancer.[65,73,74] Several reviews have noted that articles published before 1990 more often report a positive association.[19,75] This decline is likely due to recent changes in production methods and a decrease in tobacco-specific nitrosamines (TSNAs), particularly in products manufactured and sold in the United States and Europe.[76]

Oral and pharyngeal

The association with oral and pharyngeal cancers in the United States is unclear. Several studies found no or only weak positive correlation.[65,74] Boffetta et al.[73] found an increased risk, but several studies cited in this review did not control for dual-use of cigarettes and included studies published before 1990. These older studies could be measuring products that contain more carcinogens.[75] Studies with a large sample and more robust controls for confounding factors are needed to fully understand the risk. There is some evidence that the duration of use increases risk, so well-designed longitudinal studies would be beneficial.[77]

In Scandinavia, no significant association between ST, including snus, and oral or pharyngeal cancer has been found.[65,73,78–80] A longitudinal study of male construction workers showed no increased risk for either oral or pharyngeal cancer even in daily snus users who had never smoked. However, a combined oropharyngeal category did show a slight increase for this population.[78] There is some indication that an increased duration of snus use can increase the risk of cancer of the gums and buccal mucosa (the lining of the cheek and the bottom of the mouth).[73]

In South-East, an increased risk of oral and pharyngeal cancers has been demonstrated in multiple studies of ST users.[19,65,73,81] The types of ST available in these countries are higher in TSNAs and are often mixed with other substances which are known carcinogens.[19] In India, ST is regularly used with betel quid and areca nut, both known carcinogens.[65,73] However, Indian users who do not mix ST with either betel quid or areca nut are still 10 times more likely to develop these cancers than are people who do not use ST. Users in India also have an increased risk of precancerous lesions of the oral cavity.[73] ST products there are stored without refrigeration, which leads to additional fermentation that could be a causal factor.[19] For South-East Asia as a whole, the odds of developing oral cancer were 4.7 times higher for those who used ST products than for those who did not.[81]

Esophageal

The International Agency for Research on Cancer (IARC) determined that ST contains carcinogens that could result in esophageal cancer,[82] but further research is needed, as the current limited research is divided. In the United States, an increased risk has been shown. ST users who had never smoked were five times more likely to develop esophageal cancer than non-ST users.[73] In Scandinavian countries, which use predominately snus, no clear association has been shown.[73,83] In India and other South-East Asian countries, a significant positive association has been shown across several studies.[19,83] There has also been a differential risk shown by the type of ST. For example, those who use betel quid were between 2.2 and 7.1 times more likely to develop esophageal cancer.[84] The increased risk seen in South-East Asian countries but not in Scandinavian countries is likely due to the types of ST used. Though there are few studies that focus on users in the United States, it is likely that the risk is higher than for nonusers.

The predominate form of nitrosonornicotine (NNN) used in ST in the United States has been shown in laboratory experiments to cause esophageal tumors in rats.[85] In a case study of a heavy ST user, researchers found an increase in the enzyme P450, a known carcinogen,[82] in the mucus tissues of the mouth and throat, including the esophagus. These studies provide evidence for an increased risk of esophageal cancer in ST users in the United States, which is in line with Boffetta et al.'s[73] findings. However, a longitudinal study of users found no significant increase in mortality due to esophageal cancer.[67]

Pancreatic

Evidence of a link to pancreatic cancer is weak.[65] Sponsiello-Wang et al.[86] found a small association in a review of articles from the United States and Sweden, but that relationship was weakened further when controlling for cigarette use. There was a slight increase in risk in a study of Norwegian men, even among those who had never smoked. The relative risk did increase as smoking status, either former or current, was considered.[87] Higher exposure also increased the risk in Sweden and Norway.[73] While the evidence is mixed in the United States and Scandinavia, a review conducted in 2016 found no published research on the question in India.[19] This lack of research creates a large hole in the literature which needs to be filled particularly because pancreatic cancer is so deadly.

Cardiovascular disease

Several cardiovascular factors have been examined in relation to ST use, with mixed results. An acute increase in blood pressure and heart rate has been observed in male users compared to nontobacco users in western countries.[88–92] In India, a single study observed no significant difference in blood pressure between smokers and ST users, suggesting that the differences in ST products may have a differential affect.[93] This acute increase in blood pressure and heart rate may be a sign that ST increases cardiac output, which could have more serious consequences. There is also no evidence of chronic blood pressure problems.[94]

Studies in the United States and Scandinavia have found evidence of increased acute MI, fatal MI, and stroke among ST users.[94–97] However, other studies in Sweden found no association.[98–100] When associations are found, they were generally weak and occurred only when comparing ST users to smokers.[88,89,94,97,98]

In Bangladesh, the risk of chronic heart disease was increased only by one form of ST: gul. It has a higher nicotine content than other forms in Bangladesh, which could contribute to the effect. For example, gul was measured to have a nicotine concentration of 5.8% while jarda and sadapata had concentrations of 0.96% and 1.97%, respectively. This association was stronger for people with a history of long and heavy use. This is consistent with evidence that the duration and intensity of ST use increases the risk of negative cardiovascular outcomes in studies of western tobacco use.[95] However, the association was not mitigated by current gul use status. The association remained for those who had quit using gul for at least 2 years or had ever used it.[101]

Overall, the association is stronger for fatal MI and stroke than for acute problems.[88,95] ST may impede recovery from cardiovascular illnesses, contributing to a higher mortality rate.[95] Several studies found a positive association with overall morbidity, even if they did not find an increased risk of cardiovascular problems.[88,96] An important caveat is that most of these studies were done only with males, or with a limited number of females, and their results may not be applicable to women. One study from India found a positive association with chronic heart disease but only for a small sample of women.[101] Additional studies should be done with larger samples of both men and women to resolve the discrepancies.

Pregnancy outcomes

There are few studies on pregnancy outcomes, most likely because of the small proportion of women who use ST products. However, a small but consistent body of work has been done in Sweden. Snuff use in Sweden during pregnancy has been associated with an increased risk of low birth weight, preterm delivery, preeclampsia, stillbirths, and infants who are small for gestational age (SGA).[12,102–105] Users have a higher risk of delivering low birth weight infants, but not as great as that of women who smoke during pregnancy.[102,106] This increase was also seen in Alaskan natives.[107] Relatedly, Swedish women who used snuff early in pregnancy were more likely to deliver SGA infants. This association was stronger for preterm births. If use was stopped during early pregnancy, at or before 15 weeks gestation, the association disappeared.[105]

The association between ST use and preterm birth was noted in several studies, but the magnitude was not consistent.[106] England et al.[102] found that the risk was as high as that for smokers. However, Wikstrom et al.[12] found that the association was not as strong in ST users as in smokers. They did find a significant increase in the risk of both very (birth at less than 32 weeks) and moderately (birth between 32 and 36 weeks) preterm. Other studies did not examine stages of preterm birth.[12] There was no association with preterm birth among Alaskan natives, but this study had a small sample size.[108]

An increased risk of preeclampsia was found in several studies.[102,103] It was higher in ST users than in smokers.[102] There is some evidence that use early in pregnancy increases the risk of preeclampsia.[103]

There is consistent evidence for an increase in stillbirths.[106,109,110] A study in Mumbai found that 8.9% of users suffered a stillbirth as compared to 3.1% of nonusers. ST users and cigarette users had similar elevated risk. However, ST appeared to increase the risk of stillbirth earlier in pregnancy.[110]

There is some evidence linking ST to other pregnancy and neonatal complications. Gunnerbeck et al.[104] found an association with neonatal apnea, and it was stronger than the association for smoking. Studies have also found links to anemia and placental pathology in pregnant ST users, but the impact is unknown.[109]

Unfortunately, there is little information on the impacts of ST use during pregnancy on mothers or infants. Most of the existing data is from Sweden, which ignores other at-risk areas of the world.[109] From the current limited findings, it is clear that ST is not safe to use during pregnancy, though it may be recommended in some forms as a replacement for smoking in other circumstances.[12,104,109]

Conclusions and research opportunities

ST refers to a varied set of noncombustible tobacco products that are widely used in the United States and internationally. Exploring reported ST prevalences can be difficult, due to definitional differences, varying forms of ST, and the range of time periods and age groups included in each study. Table 3.3 gives an overview of several large studies reviewed in this chapter, including factors (e.g., product type, period of reporting) that may impact the prevalence rate when compared to other studies.

Table 3.3 Prevalence of smokeless tobacco use across countries and surveys.

Country	Survey	Product type reported	Period of reported use	Survey Year(s)	Age group	Prevalence (%)
United States	NATS	SLT	Current	2012	18+	1.7
United States	MTF	SLT	Past 30 days	2017	Grades 8, 10, and 12	1.7–4.8
United States	NYTS	SLT	Past 30 days	2017	Grades 6–12	1.9–5.5
United States	TPRPS	SLT	Past 30 days	2014–16	18+	2.3–3.7
United States	NSDUH	SLT	Past 30 days	2014	12+	3.3
Europe	PPACTE	SLT	Current	2010	15+	1.7
Europe	SE385	SLT	Current	2012	15+	2.0
India	GATS	Khaini	Current	2009	15+	11.6
India	GATS	SLT	Current	2009	15+	25.9
Sweden	PPACTE	Snus	Regular use	2010	15+	12.3
Sweden	GATS/ SE385	SLT	Current	2008–12	15+	17.7
Bangladesh	GATS	SLT	Current	2004	15+	19.7
Bangladesh	GATS	SLT	Current	2009	15+	27.2

GATS, Global Adult Tobacco Survey; *MTF*, Monitoring the Future; *NATS*, National Adult Tobacco Survey; *NSDUH*, National Survey on Drug Use and Health; *NYTS*, National Youth Tobacco Survey; *PATH*, Population Assessment of Tobacco and Health Study; *PPACTE*, Pricing Policy and Control of Tobacco in Europe; *SE385*, Special Eurobarometer 385; *TPRPSs*, Tobacco Products Risk Perception Surveys; *YSS*, Youth Smoking Survey.

Though prevalences vary widely across studies, there are several uniting factors and take-home messages. ST use internationally is more common than in the United States. Use in South-East Asia is particularly high. More men than women use in the United States. This is also true in most of the world but use in some South-East Asian countries is higher among females than males. Generally, the rates of use are higher in young adults than in either adolescents or older adults, but adolescence is a period of heighted risk for initiation. Flavoring is common in ST products, which make them more appealing to youths and first-time users. Concurrent use of more than one tobacco product (poly- or dual-use) is fairly common among ST users.

There is some argument that the negative health outcomes of ST are less damaging than those of cigarettes. However, this is only applicable to the types of ST manufactured and sold in the United States and Europe. The ST products used in South-East Asia are often mixed with other substances which contain more carcinogens than ST on its own. ST products have been linked to an increased likelihood of dental disease; oral, pharyngeal, esophageal, and pancreatic cancers; negative pregnancy outcomes; and overall mortality. Health outcomes vary widely internationally due to the variety of products and manufacturing practices.

Research on ST use and its health outcomes is relatively well developed, but there are several areas where more work is needed. One key weakness in the existing literature is that most studies are cross-sectional. Well-designed longitudinal studies could provide valuable insight into long-term health outcomes, the impact of intensity of use, and patterns of use and poly-use over the life course. Cross-sectional research can give us some information about these topics, but longitudinal research will enable researchers to come to more developed conclusions and potentially open up new questions as tobacco use changes around the world. PATH is a particularly valuable cohort study whose data could be leveraged for advancing our understanding of the trajectories of ST use in the United States.

There are also several subpopulations in which the use of ST should be better explored. For example, there are several papers[111,112] on United States military members. There is a focus on Air Force members in this research,[113] but a wider look at military personnel would be beneficial. Another subpopulation of special interest is sexual minorities. Research on sexual minorities (i.e., those who do not identify as heterosexual) has shown that this group is more likely to have mental health and substance use issues, including cigarette use, than heterosexual people within the same age group.[7,114] Some work has been done on sexual minorities and ST use,[115] but a deeper exploration into this population could contribute to a better understanding of the interaction between sexual orientation and substance use. This research should also be expanded to include gender minorities, i.e., those that do not identify as cisgender.

There are also opportunities to expand our knowledge of risk factors and etiology of ST use. Work has been done on the genetic etiology of substance use disorders generally.[116] More specifically, Wilkinson et al.[117] found evidence for a genetic link to ST use among Mexican heritage youths. To fully explore this interaction, a well-developed study design, which includes both a survey component and possible biological collection, would be ideal. Further examining genetics as a risk factor for use may enable researchers to better understand the etiology of tobacco use and substance use as a whole.

References

1. Organization WH. *WHO Framework Convention on Tobacco Control.* Geneva: World Health Organization; 2003.
2. Melikian AA, Hoffmann D. Smokeless tobacco: a gateway to smoking or a way away from smoking. *Biomarkers.* 2009;14(sup1):85–89.
3. Howard-Pitney B, Winkleby MA. Chewing tobacco: who uses and who quits? Findings from NHANES III, 1988–1994. National health and nutrition examination survey III. *Am J Public Health.* 2002;92(2):250–256.
4. Agaku IT, Vardavas CI, Ayo-Yusuf OA, Alpert HR, Connolly GN. Temporal trends in smokeless tobacco use among US middle and high school students, 2000–2011. *JAMA.* 2013;309(19):1992–1994.
5. Maher JE, Bushore CJ, Rohde K, Dent CW, Peterson E. Is smokeless tobacco use becoming more common among US male smokers? Trends in Alaska. *Addict Behav.* 2012;37(7):862–865.
6. Lee YO, Hebert CJ, Nonnemaker JM, Kim AE. Multiple tobacco product use among adults in the United States: cigarettes, cigars, electronic cigarettes, hookah, smokeless tobacco, and snus. *Prev Med.* 2014;62:14–19.
7. Kasza KA, Ambrose BK, Conway KP, et al. Tobacco-product use by adults and youths in the United States in 2013 and 2014. *N Engl J Med.* 2017;376(4):342–353.
8. Jones DM, Majeed BA, Weaver SR, Sterling K, Pechacek TF, Eriksen MP. Prevalence and factors associated with smokeless tobacco use, 2014–2016. *Am J Health Behav.* 2017;41(5):608–617.
9. Lipari RN, Van Horn SL. Trends in smokeless tobacco use and initiation: 2002 to 2014. *The CBHSQ Report: May 31, 2017. Center for Behavioral Health Statistics and Quality, Substance Abuse and Mental Health Services Administration, Rockville, MD.* 2013.
10. Lund KE, McNeill A. Patterns of dual use of snus and cigarettes in a mature snus market. *Nicotine Tob Res.* 2012;15(3):678–684.
11. Leon ME, Lugo A, Boffetta P, et al. Smokeless tobacco use in Sweden and other 17 European countries. *Eur J Public Health.* 2016;26(5):817–821.

12. Wikstrom AK, Cnattingius S, Galanti MR, Kieler H, Stephansson O. Effect of Swedish snuff (snus) on preterm birth. *BJOG An Int J Obstet Gynaecol.* 2010;117(8): 1005–1010.
13. Norberg M, Malmberg G, Ng N, Brostrom G. Who is using snus? - time trends, socioeconomic and geographic characteristics of snus users in the ageing Swedish population. *BMC Public Health.* 2011;11:929.
14. Eurobarometer S. *385. Attitudes of Europeans towards Tobacco.* 2012.
15. Agaku IT, Filippidis FT, Vardavas CI, et al. Poly-tobacco use among adults in 44 countries during 2008–2012: evidence for an integrative and comprehensive approach in tobacco control. *Drug Alcohol Depend.* 2014;139:60–70.
16. Sinha DN, Agarwal N, Gupta P. Prevalence of smokeless tobacco use and number of users in 121 countries. *Br J Med Med Res.* 2015;9(6):1–20.
17. Sinha D, Gupta P, Ray C, Singh P. Prevalence of smokeless tobacco use among adults in WHO South-East Asia. *Indian J Cancer.* 2012;49(4):342.
18. Abdullah AS, Driezen P, Ruthbah UH, Nargis N, Quah AC, Fong GT. Patterns and predictors of smokeless tobacco use among adults in Bangladesh: findings from the International Tobacco Control (ITC) Bangladesh Survey. *PLoS One.* 2014;9(7): e101934.
19. Sinha DN, Abdulkader RS, Gupta PC. Smokeless tobacco-associated cancers: a systematic review and meta-analysis of Indian studies. *Int J Cancer.* 2016;138(6): 1368–1379.
20. Sinha DN, Suliankatchi RA, Gupta PC, et al. Global burden of all-cause and cause-specific mortality due to smokeless tobacco use: systematic review and meta-analysis. *Tob Control.* 2018;27(1):35–42.
21. Sreeramareddy CT, Pradhan PMS, Mir IA, Sin S. Smoking and smokeless tobacco use in nine South and Southeast Asian countries: prevalence estimates and social determinants from demographic and health surveys. *Popul Health Metrics.* 2014;12(1):22.
22. Prabhakar V, Jayakrishnan G, Nair S, Ranganathan B. Determination of trace metals, moisture, pH and assessment of potential toxicity of selected smokeless tobacco products. *Indian J Pharm Sci.* 2013;75(3):262.
23. Kostova D, Dave D. Smokeless tobacco use in India: role of prices and advertising. *Soc Sci Med.* 2015;138:82–90.
24. Klesges RC, Ebbert JO, Morgan GD, et al. Impact of differing definitions of dual tobacco use: implications for studying dual use and a call for operational definitions. *Nicotine Tob Res.* 2011;13(7):523–531.
25. Rodu B, Plurphanswat N. E-cigarette use among US adults: population assessment of tobacco and health (PATH) study. *Nicotine Tob Res.* 2017;20(8):940–948.
26. Bombard JM, Pederson LL, Nelson DE, Malarcher AM. Are smokers only using cigarettes? Exploring current polytobacco use among an adult population. *Addict Behav.* 2007;32(10):2411–2419.
27. Wetter DW, McClure JB, de Moor C, et al. Concomitant use of cigarettes and smokeless tobacco: prevalence, correlates, and predictors of tobacco cessation. *Prev Med.* 2002;34(6):638–648.
28. Little MA, Bursac Z, Derefinko KJ, et al. Types of dual and poly-tobacco users in the US military. *Am J Epidemiol.* 2016;184(3):211–218.
29. National Cancer Institute. (n.d.). 2010–11 TUS-CPS Data, Tables 4/4b: Use of Cigars & Smokeless Tobacco Products; Use of Pipes & Hookah Tobacco Products. Retrieved from https://cancercontrol.cancer.gov/brp/tcrb/tus-cps/results/data1011/table4.html.
30. Fix BV, O'Connor RJ, Vogl L, et al. Patterns and correlates of polytobacco use in the United States over a decade: NSDUH 2002–2011. *Addict Behav.* 2014;39(4): 768–781.

31. McClave-Regan AK, Berkowitz J. Smokers who are also using smokeless tobacco products in the US: a national assessment of characteristics, behaviours and beliefs of 'dual users'. *Tob Control.* 2011;20(3):239–242.
32. Tomar SL, Alpert HR, Connolly GN. Patterns of dual use of cigarettes and smokeless tobacco among US males: findings from national surveys. *Tob Control.* 2010;19(2): 104–109.
33. Osibogun O, Taleb ZB, Bahelah R, Salloum RG, Maziak W. Correlates of poly-tobacco use among youth and young adults: findings from the Population Assessment of Tobacco and Health study, 2013–2014. *Drug Alcohol Depend.* 2018;187:160–164.
34. Mushtaq N, Williams MB, Beebe LA. Concurrent use of cigarettes and smokeless tobacco among US males and females. *J Environ Public Health.* 2012;2012:984561.
35. Sung HY, Wang Y, Yao T, Lightwood J, Max W. Polytobacco use of cigarettes, cigars, chewing tobacco, and snuff among US adults. *Nicotine Tob Res.* 2016;18(5): 817–826.
36. Creamer MR, Portillo GV, Clendennen SL, Perry CL. Is adolescent poly-tobacco use associated with alcohol and other drug use? *Am J Health Behav.* 2016;40(1):117–122.
37. Teo KK, Ounpuu S, Hawken S, et al. Tobacco use and risk of myocardial infarction in 52 countries in the INTERHEART study: a case-control study. *Lancet.* 2006; 368(9536):647–658.
38. Delnevo CD, Wackowski OA, Giovenco DP, Manderski MTB, Hrywna M, Ling PM. Examining market trends in the United States smokeless tobacco use: 2005–2011. *Tob Control.* 2014;23(2):107–112.
39. Huang L-L, Baker HM, Meernik C, Ranney LM, Richardson A, Goldstein AO. Impact of non-menthol flavours in tobacco products on perceptions and use among youth, young adults and adults: a systematic review. *Tob Control.* 2017;26(6):709–719.
40. Villanti AC, Johnson AL, Ambrose BK, et al. Flavored tobacco product use in youth and adults: findings from the first wave of the PATH study (2013–2014). *Am J Prev Med.* 2017;53(2):139–151.
41. Dai H. Peer reviewed: single, dual, and poly use of flavored tobacco products among youths. *Prev Chronic Dis.* 2018;15.
42. Ambrose BK, Day HR, Rostron B, et al. Flavored tobacco product use among US youth aged 12–17 years, 2013–2014. *JAMA.* 2015;314(17):1871–1873.
43. Oliver AJ, Jensen JA, Vogel RI, Anderson AJ, Hatsukami DK. Flavored and nonflavored smokeless tobacco products: rate, pattern of use, and effects. *Nicotine Tob Res.* 2012;15(1):88–92.
44. Chaffee BW, Urata J, Couch ET, Gansky SA. Perceived flavored smokeless tobacco ease-of-use and youth susceptibility. *Tob Regul Sci.* 2017;3(3):367–373.
45. Kennedy RD, Leatherdale ST, Burkhalter R, Ahmed R. Prevalence of smokeless tobacco use among Canadian youth between 2004 and 2008: findings from the Youth Smoking Survey. *Can J Public Health.* 2011:358–363.
46. Chang JT, Levy DT, Meza R. Trends and factors related to smokeless tobacco use in the United States. *Nicotine Tob Res.* 2016;18(8):1740–1748.
47. Cheng Y-C, Rostron BL, Day HR, et al. Patterns of use of smokeless tobacco in US adults, 2013–2014. *Am J Public Health.* 2017;107(9):1508–1514.
48. McClave A, Rock V, Thorne S, Malarcher A. State-specific prevalence of cigarette smoking and smokeless tobacco use among adults-United States, 2009. *MMWR (Morb Mortal Wkly Rep).* 2010;59(43):1400–1406.
49. Smith ML, Colwell B, Forté CA, Pulczinski JC, McKyer ELJ. Psychosocial correlates of smokeless tobacco use among Indiana adolescents. *J Community Health.* 2015;40(2): 208–214.

50. Johnston LD, Miech RA, O'Malley PM, Bachman JG, Schulenberg JE, Patrick ME. *Monitoring the Future National Survey Results on Drug Use, 1975–2017: Overview, Key Findings on Adolescent Drug Use*. 2018.
51. Wang TW, Gentzke A, Sharapova S, Cullen KA, Ambrose BK, Jamal A. Tobacco product use among middle and high school students—United States, 2011–2017. *MMWR (Morb Mortal Wkly Rep)*. 2018;67(22):629.
52. Holman LR, Bricker JB, Comstock BA. Psychological predictors of male smokeless tobacco use initiation and cessation: a 16-year longitudinal study. *Addiction*. 2013; 108(7):1327–1335.
53. White TJ, Redner R, Bunn JY, Higgins ST. Do socioeconomic risk factors for cigarette smoking extend to smokeless tobacco use? *Nicotine Tob Res*. 2016;18(5): 869–873.
54. Noonan D, Duffy SA. Factors associated with smokeless tobacco use and dual use among blue collar workers. *Public Health Nurs*. 2014;31(1):19–27.
55. Jitnarin N, Haddock CK, Poston WS, Jahnke S. Smokeless tobacco and dual use among firefighters in the central United States. *J Environ Public Health*. 2013;2013: 675426.
56. Couch ET, Darius E, Walsh MM, Chaffee BW. Smokeless tobacco decision-making among rural adolescent males in California. *J Community Health*. 2017;42(3):544–550.
57. Nemeth JM, Liu ST, Klein EG, Ferketich AK, Kwan M-P, Wewers ME. Factors influencing smokeless tobacco use in rural Ohio Appalachia. *J Community Health*. 2012;37(6):1208–1217.
58. Conway KP, Green VR, Kasza KA, et al. Co-occurrence of tobacco product use, substance use, and mental health problems among adults: findings from Wave 1 (2013–2014) of the Population Assessment of Tobacco and Health (PATH) Study. *Drug Alcohol Depend*. 2017;177:104–111.
59. Fu Q, Vaughn MG, Wu LT, Heath AC. Psychiatric correlates of snuff and chewing tobacco use. *PLoS One*. 2014;9(12):e113196.
60. Hermes ED, Wells TS, Smith B, et al. Smokeless tobacco use related to military deployment, cigarettes and mental health symptoms in a large, prospective cohort study among US service members. *Addiction*. 2012;107(5):983–994.
61. Sawchuk CN, Roy-Byrne P, Noonan C, et al. Smokeless tobacco use and its relation to panic disorder, major depression, and posttraumatic stress disorder in American Indians. *Nicotine Tob Res*. 2012;14(9):1048–1056.
62. Engstrom K, Magnusson C, Galanti MR. Socio-demographic, lifestyle and health characteristics among snus users and dual tobacco users in Stockholm County, Sweden. *BMC Public Health*. 2010;10:619.
63. Grotvedt L, Stigum H, Hovengen R, Graff-Iversen S. Social differences in smoking and snuff use among Norwegian adolescents: a population based survey. *BMC Public Health*. 2008;8:322.
64. Øverland S, Tjora T, Hetland J, Aaro LE. Associations between adolescent socioeducational status and use of snus and smoking. *Tob Control*. 2010;19(4):291–296.
65. Critchley JA, Unal B. Health effects associated with smokeless tobacco: a systematic review. *Thorax*. 2003;58(5):435–443.
66. Siddiqi K, Shah S, Abbas SM, et al. Global burden of disease due to smokeless tobacco consumption in adults: analysis of data from 113 countries. *BMC Med*. 2015;13:194.
67. Timberlake DS, Nikitin D, Johnson NJ, Altekruse SF. A longitudinal study of smokeless tobacco use and mortality in the United States. *Int J Cancer*. 2017;141(2):264–270.
68. Phillips CV, Heavner KK. Smokeless tobacco: the epidemiology and politics of harm. *Biomarkers*. 2009;14(sup1):79–84.
69. Lund KE. Association between willingness to use snus to quit smoking and perception of relative risk between snus and cigarettes. *Nicotine Tob Res*. 2012;14(10):1221–1228.

70. Lee PN. Epidemiological evidence relating snus to health—an updated review based on recent publications. *Harm Reduct J.* 2013;10:36.
71. Fisher MA, Taylor GW, Tilashalski KR. Smokeless tobacco and severe active periodontal disease, NHANES III. *J Dent Res.* 2005;84(8):705—710.
72. Chu YH, Tatakis DN, Wee AG. Smokeless tobacco use and periodontal health in a rural male population. *J Periodontol.* 2010;81(6):848—854.
73. Boffetta P, Hecht S, Gray N, Gupta P, Straif K. Smokeless tobacco and cancer. *Lancet Oncol.* 2008;9(7):667—675.
74. Lee PN, Hamling J. The relation between smokeless tobacco and cancer in Northern Europe and North America. A commentary on differences between the conclusions reached by two recent reviews. *BMC Cancer.* 2009;9:256.
75. Lee PN, Hamling J. Systematic review of the relation between smokeless tobacco and cancer in Europe and North America. *BMC Med.* 2009;7:36.
76. Janbaz KH, Qadir MI, Basser HT, Bokhari TH, Ahmad B. Risk for oral cancer from smokeless tobacco. *Contemp Oncol.* 2014;18(3):160—164.
77. Zhou JC, Michaud DS, Langevin SM, McClean MD, Eliot M, Kelsey KT. Smokeless tobacco and risk of head and neck cancer: evidence from a case-control study in New England. *Int J Cancer.* 2013;132(8):1911—1917.
78. Luo J, Ye W, Zendehdel K, et al. Oral use of Swedish moist snuff (snus) and risk for cancer of the mouth, lung, and pancreas in male construction workers: a retrospective cohort study. *Lancet.* 2007;369(9578):2015—2020.
79. Roosaar A, Johansson AL, Sandborgh-Englund G, Axell T, Nyren O. Cancer and mortality among users and nonusers of snus. *Int J Cancer.* 2008;123(1):168—173.
80. Weitkunat R, Sanders E, Lee PN. Meta-analysis of the relation between European and American smokeless tobacco and oral cancer. *BMC Public Health.* 2007;7.
81. Khan Z, Tonnies J, Muller S. Smokeless tobacco and oral cancer in South Asia: a systematic review with meta-analysis. *J Cancer Epidemiol.* 2014;2014:394696.
82. Mallery SR, Tong M, Michaels GC, Kiyani AR, Hecht SS. Clinical and biochemical studies support smokeless tobacco's carcinogenic potential in the human oral cavity. *Cancer Prev Res.* 2014;7(1):23—32.
83. Gupta R, Gupta S, Sharma S, Sinha DN, Mehrotra R. Risk of coronary heart disease among smokeless tobacco users: results of systematic review and meta-analysis of global data. *Nicotine Tob Res.* 2018;2018. https://doi.org/10.1093/ntr/nty002 (Epub ahead of print).
84. Gupta PC, Ray CS, Sinha DN, Singh PK. Smokeless tobacco: a major public health problem in the SEA region: a review. *Indian J Public Health.* 2011;55(3):199.
85. Balbo S, James-Yi S, Johnson CS, et al. (S)-N′-Nitrosonornicotine, a constituent of smokeless tobacco, is a powerful oral cavity carcinogen in rats. *Carcinogenesis.* 2013; 34(9):2178—2183.
86. Sponsiello-Wang Z, Weitkunat R, Lee PN. Systematic review of the relation between smokeless tobacco and cancer of the pancreas in Europe and North America. *BMC Cancer.* 2008;8.
87. Boffetta P, Aagnes B, Weiderpass E, Andersen A. Smokeless tobacco use and risk of cancer of the pancreas and other organs. *Int J Cancer.* 2005;114(6):992—995.
88. Asplund K. Smokeless tobacco and cardiovascular disease. *Prog Cardiovasc Dis.* 2003; 45(5):383—394.
89. Gupta R, Gurm H, Bartholomew JR. Smokeless tobacco and cardiovascular risk. *Arch Intern Med.* 2004;164(17):1845—1849.
90. Wolk R, Shamsuzzaman AS, Svatikova A, et al. Hemodynamic and autonomic effects of smokeless tobacco in healthy young men. *J Am Coll Cardiol.* 2005;45(6):910—914.
91. Hergens MP, Lambe M, Pershagen G, Ye W. Risk of hypertension amongst Swedish male snuff users: a prospective study. *J Intern Med.* 2008;264(2):187—194.

92. Øverland S, Skogen JC, Lissner L, Bjerkeset O, Tjora T, Stewart R. Snus use and cardiovascular risk factors in the general population: the HUNT3 study. *Addiction*. 2013; 108(11):2019–2028.
93. Anand A, MIK S. The risk of hypertension and other chronic diseases: comparing smokeless tobacco with smoking. *Front public health*. 2017;5:255.
94. Lee PN. Circulatory disease and smokeless tobacco in Western populations: a review of the evidence. *Int J Epidemiol*. 2007;36(4):789–804.
95. Piano MR, Benowitz NL, Fitzgerald GA, et al. Impact of smokeless tobacco products on cardiovascular disease: implications for policy, prevention, and treatment: a policy statement from the American Heart Association. *Circulation*. 2010;122(15): 1520–1544.
96. Hergens MP, Alfredsson L, Bolinder G, Lambe M, Pershagen G, Ye W. Long-term use of Swedish moist snuff and the risk of myocardial infarction amongst men. *J Intern Med*. 2007;262(3):351–359.
97. Boffetta P, Straif K. Use of smokeless tobacco and risk of myocardial infarction and stroke: systematic review with meta-analysis. *BMJ*. 2009;339:b3060.
98. Hansson J, Pedersen NL, Galanti MR, et al. Use of snus and risk for cardiovascular disease: results from the Swedish Twin Registry. *J Intern Med*. 2009;265(6):717–724.
99. Janzon E, Hedblad B. Swedish snuff and incidence of cardiovascular disease. A population-based cohort study. *BMC Cardiovasc Disord*. 2009;9:21.
100. Hansson J, Galanti MR, Hergens MP, et al. Snus (Swedish smokeless tobacco) use and risk of stroke: pooled analyses of incidence and survival. *J Intern Med*. 2014;276(1): 87–95.
101. Rahman MA, Spurrier N, Mahmood MA, Rahman M, Choudhury SR, Leeder S. Is there any association between use of smokeless tobacco products and coronary heart disease in Bangladesh? *PLoS One*. 2012;7(1):e30584.
102. England LJ, Levine RJ, Mills JL, Klebanoff MA, Yu KF, Cnattingius S. Adverse pregnancy outcomes in snuff users. *Am J Obstet Gynecol*. 2003;189(4):939–943.
103. Wikstrom AK, Stephansson O, Cnattingius S. Tobacco use during pregnancy and preeclampsia risk: effects of cigarette smoking and snuff. *Hypertension*. 2010;55(5): 1254–1259.
104. Gunnerbeck A, Wikstrom AK, Bonamy AK, Wickstrom R, Cnattingius S. Relationship of maternal snuff use and cigarette smoking with neonatal apnea. *Pediatrics*. 2011; 128(3):503–509.
105. Baba S, Wikstrom AK, Stephansson O, Cnattingius S. Changes in snuff and smoking habits in Swedish pregnant women and risk for small for gestational age births. *BJOG An Int J Obstet Gynaecol*. 2013;120(4):456–462.
106. Inamdar AS, Croucher RE, Chokhandre MK, Mashyakhy MH, Marinho VC. Maternal smokeless tobacco use in pregnancy and adverse health outcomes in newborns: a systematic review. *Nicotine Tob Res*. 2015;17(9):1058–1066.
107. England LJ, Kim SY, Shapiro-Mendoza CK, et al. Maternal smokeless tobacco use in Alaska Native women and singleton infant birth size. *Acta Obstet Gynecol Scand*. 2012; 91(1):93–103.
108. England LJ, Kim SY, Shapiro-Mendoza CK, et al. Effects of maternal smokeless tobacco use on selected pregnancy outcomes in Alaska Native women: a case–control study. *Acta Obstet Gynecol Scand*. 2013;92(6):648–655.
109. England LJ, Kim SY, Tomar SL, et al. Non-cigarette tobacco use among women and adverse pregnancy outcomes. *Acta Obstet Gynecol Scand*. 2010;89(4):454–464.
110. Gupta PC, Subramoney S. Smokeless tobacco use and risk of stillbirth: a cohort study in Mumbai, India. *Epidemiology*. 2006;17(1):47–51.
111. Bergman HE, Hunt YM, Augustson E. Smokeless tobacco use in the United States military: a systematic review. *Nicotine Tob Res*. 2011;14(5):507–515.

112. Peterson AL, Severson HH, Andrews JA, et al. Smokeless tobacco use in military personnel. *Mil Med*. 2007;172(12):1300–1305.
113. Dunkle A, Kalpinski R, Ebbert J, Talcott W, Klesges R, Little MA. Predicting smokeless tobacco initiation and re-initiation in the United States Air Force. *Addict Behav Rep*. 2018;96(5):897–905.
114. Medley G, Lipari RN, Bose J, Cribb DS, Kroutil LA, McHenry G. *Sexual Orientation and Estimates of Adult Substance Use and Mental Health: Results from the 2015 National Survey on Drug Use and Health*. 2016. NSDUH Data Review. Retrieved from; http://www.samhsa.gov/data/.
115. Johnson SE, Holder-Hayes E, Tessman GK, King BA, Alexander T, Zhao X. Tobacco product use among sexual minority adults: findings from the 2012–2013 national adult tobacco survey. *Am J Prev Med*. 2016;50(4):e91–e100.
116. Merikangas KR, Conway KP. Genetic epidemiology of substance use disorders. In: Charney DS, Nestler EJ, eds. *Neurobiology of Mental Illness*. New York, NY: Oxford University Press; 2013.
117. Wilkinson AV, Koehly LM, Vandewater EA, et al. Demographic, psychosocial, and genetic risk associated with smokeless tobacco use among Mexican heritage youth. *BMC Medical Genetics*. 2015;16(1):43.

CHAPTER FOUR

Clinical laboratory studies of smokeless tobacco use

Bartosz Koszowski[1], Wallace B. Pickworth[1], Steven E. Meredith[2], Lynn C. Hull[2]
[1]Battelle Memorial Institute, Baltimore, MD, United States
[2]U.S. Food and Drug Administration, Silver Spring, MD, United States

Value of clinical laboratory studies of smokeless tobacco

Nicotine is the primary addictive constituent of tobacco products.[1–4] As with all drugs of abuse, the dose of nicotine is associated with abuse liability. This is the case whether nicotine is administered alone (e.g., intravenously or by nasal spray in a laboratory setting) or in a tobacco product.[5,6]

Exposure to nicotine from tobacco products, including smokeless tobacco (ST), is dependent on a variety of factors. One study found that ST use topography (e.g., duration of use, frequency of use, amount used) is associated with nicotine exposure.[7] Another study found that consuming coffee and cola may reduce nicotine exposure from nicotine gum, potentially due to the acidification of the mouth and the effect of pH on nicotine absorption,[8] a result that could generalize to ST. In addition, characteristics of ST products, including nicotine content, pH, and other ingredients, can affect nicotine exposure.[9] Considerable evidence suggests that changes in the free nicotine content of ST products independently affect nicotine exposure and abuse liability.

Results from laboratory-based clinical studies of tobacco products have led to better understanding of use patterns, toxicant exposure, and ultimately health risks. Studies by Henningfield and colleagues at the National Institute on Drug Abuse (NIDA) Addiction Research Center (later the Intramural Research Program) provided scientific evidence that nicotine was an addictive substance and that tobacco products delivered nicotine.[5,10–12] These findings, along with the abundance of research that followed, have had important social, public health, and regulatory implications resulting in reductions in youth access to tobacco, adult cigarette smoking prevalence,

Smokeless Tobacco Products
ISBN: 978-0-12-818158-4
https://doi.org/10.1016/B978-0-12-818158-4.00004-2

and smoking-related disease. However, ST use prevalence has been resistant to reduction, especially among the most at-risk groups such as youth and adult non-Hispanic males, including those in Native American/Alaskan and rural populations.[13]

The purpose of this chapter is to review the findings of clinical laboratory studies of ST use characteristics, user preferences, and the effects of additives such as those that alter flavor and may affect nicotine bioavailability. Clinical studies are especially well suited for investigation of ST product characteristics because a repeated measures design provides statistically powerful results from a relatively small number of participants. This approach has long been a standard of practice for the determination of drug effects. Other experimental aspects such as the use of placebo products, comparison of study product with own product, and directed ST use can uncover or control for expectation, product familiarity, and difference in acute versus chronic use. Laboratory-based research can be designed to minimize inherent variations in product usage among the general population by controlling the amount of product and length of time used. Validated questionnaires allow understanding of the subjective effects engendered by ST consumption. Questionnaires that assess level of dependence—e.g., the Fagerström Test for Nicotine Dependence-Smokeless Tobacco (FTND-ST, Fig. 4.1) and the Hooked on Nicotine Checklist modified for ST (HONC, Fig. 4.2)—provide a measure of the baseline dependence of the study population. Level of dependence can be used as a covariate in subsequent analyses of product liking and risk perceptions. Measures of liking allow comparison of the multitude of products and flavors commercially available.

Other instruments that measure nicotine dependence are the Wisconsin Inventory of Smoking Dependence Motivations (WISDM),[14] the Nicotine Dependence Syndrome Scale (NDSS),[15] and the Heaviness of Smoking Index (HSI).[16] Historically, nicotine dependence instruments were developed to assess cigarette smokers only and were later modified for other tobacco product users. Results need to be interpreted with caution, especially when applied to populations that differ in significant ways from the original validation population, even if subjects are using the same tobacco product as the group in which the measure was validated.

Estimation of product value using behavioral economics techniques is an emerging tool for assessing tobacco product abuse liability. These tests contain choice procedure methods to evaluate relative reinforcing effects. Participants are asked to estimate an actual or hypothetical purchase price, or how much of a product they would be willing to purchase at increasing

Interviewer instructions: ask all questions, read all response options, and circle the response given.

Once complete, add up the points corresponding to each response.

INTERVIEWER: I will now ask you some specific questions about your smokeless tobacco use. We will go through the questions together and feel free to ask any questions if anything is unclear.

1. How soon after you wake up do you place your first dip? Is it within...	5 minutes........Score:3 6-30 minutes........Score:2 31-60 minutes........Score:1 After 60 minutes........Score:0
2. How often do you intentionally swallow tobacco juice? Is it...	Always........Score:2 Sometimes........Score:1 Never........Score:0
3. Which chew would you hate most to give up?	The first one in the morningScore:1 Any other........Score:0
4. How many cans/pouches per week do you use?	More than 3........Score:2 2-3........Score:1 1........Score:0
5. Do you chew more frequently during the first hours after waking than during the rest of the day?	Yes........Score:1 No........Score:0
6. Do you chew if you are so ill that you are in bed most of the day?	Yes........Score:1 No........Score:0

Figure 4.1 The Fagerström Test for Nicotine Dependence modified for ST. The FTND-ST total score is calculated as the sum of the individual items. *Adapted from Ebbert JO, Patten CA, Schroeder DR. The Fagerstrüm Test for Nicotine Dependence-Smokeless Tobacco (FTND-ST). Addict Behav. 2006;31(9):1716-1721.*

prices. Higher estimated prices, or the willingness to purchase product at increasing price points (e.g., a cigarette purchase task), are correlated with increased abuse liability. The methods can be used for comparison within an individual product class (e.g., cigarette to cigarette) or between product classes (e.g., cigarette to ST). The task can be amended to accommodate the purchase valence of a product in various contrived situations (e.g., how much a product would be worth to a participant under no unusual restrictions, during abstinence, after having food or drink, or after consumption of other drugs).

Clinical laboratory studies predict real life use behavior and the public and personal health risk associated with particular products, flavors, and packaging. Liu and colleagues performed a qualitative analysis of adolescent and adult perceptions of novel and traditional ST products in a group of participants from rural Ohio.[17] Using focus groups and qualitative interviews, they found that adolescents and adults held similar beliefs and reactions to the appearance of ST packaging. Perceptions of quality and price of

Please select the appropriate response – Yes or No.

		Yes	No
1.	Have you ever tried to quit using smokeless tobacco, but couldn't?		
2.	Do you use smokeless tobacco now because it is really hard to quit?		
3.	Have you ever felt like you were addicted to tobacco?		
4.	Do you ever have strong cravings to use smokeless tobacco?		
5.	Have you ever felt like you really needed to use smokeless tobacco?		
6.	Is it hard to keep from using smokeless tobacco in places where you are not supposed to?		
When you tried to quit using smokeless tobacco OR When you haven't used smokeless tobacco for a while....			
7.	Did you find it hard to concentrate because you couldn't use smokeless tobacco?		
8.	Did you feel irritable because you couldn't use smokeless tobacco?		
9.	Did you feel a strong urge or need to use smokeless tobacco?		
10.	Did you feel nervous, restless, or anxious because you couldn't use smokeless tobacco?		

Figure 4.2 The Hooked on Nicotine Checklist modified for ST. Scoring of the HONC consists of two methods: 1) total scale score as an indicator of level of nicotine dependence, and 2) endorsement of at least one item as an indicator of "loss of autonomy" over tobacco use. *Based on HONC originally published by DiFranza J, Savageau J, Rigotti N, Fletcher K, Ockene J, McNeill A, Coleman M, and Wood C. Development of symptoms of tobacco dependence in youths: 30 month follow up data from the DANDY study. Tob Control. 2002;11(3):228-235.*

traditional products were closely related to taste and packaging. Packaging color, design, and size or shape influenced decisions to purchase.

Hatsukami and colleagues developed the Product Evaluation Scale (PES; Fig. 4.3).[18] It has items adapted from the modified Cigarette Evaluation Questionnaire (mCEQ),[19] with other items rating the subjective effects such as sensation in the mouth, taste, and reduction of craving and withdrawal. Smokers were directed to sample a variety of noncombusted oral tobacco products, including ST, during smoking abstinence and then choose a product to substitute for cigarettes for 2 weeks.[18] The results were used to evaluate and validate the PES. The validity of the PES was shown to be supported by the product chosen for the 2-week substitution period; substitutes with higher mean ratings on satisfaction, withdrawal, and craving relief and lower ratings on aversion were chosen by more participants than other products.

Nicotine absorption

Several factors modulate nicotine absorption. Some are user specific, such as local blood flow and behavior (e.g., amount, frequency, mouth

Subjective Responses to Oral Tobacco Products: Scale Validation

1. Was it satisfying?
2. Did it taste good?
3. Did you enjoy the sensations in your mouth?
4. Did it calm you down?
5. Did it make you feel more awake?
6. Did it make you feel less irritable?
7. Did it help you concentrate?
8. Did it reduce your hunger for food?
9. Did it make you dizzy?
10. Did it make you nauseous?
11. Did it immediately relieve your craving for a cigarette?
12. Did you enjoy it?
13. Did it relieve withdrawal symptoms?
14. Did it relieve the urge to smoke?
15. Was it enough nicotine?
16. Was it too much nicotine?
17. Was it easy to use?
18. Were there bothersome side effects?
19. Were you comfortable using the product in public?
20. Did you still have a craving for a cigarette after using the product?
21. Are you concerned that you would become dependent on this product?

Notes. A) Rated on a 1–7 scale, where 1 (not at all) and 7 (extremely); subscale scores are the following: satisfaction (items 1, 2, 3, and 12); psychological reward (items 4, 5, 6, 7, and 8); aversion (items 9, 10, 16, and 18); relief (items 11, 13, 14, 15, and reversed for 20), item scores for each subscale were averaged. (Adapted from Hatsukami DK, Zhang Y, O'Connor RJ, Severson HH. *Nicotine Tob Res*. 2013:15(7):1259–1264. Published online 2012 Dec 13. doi: 10.1093/ntr/nts265.)

Figure 4.3 The Product Evaluation Scale modified for ST.

placement, and movement of bolus). Product characteristics also affect absorption, such as nicotine content, nontobacco ingredients (e.g., menthol and methyl salicylate, an ingredient used in wintergreen flavoring),[20] "wettability," tobacco cut size (which determines size and surface area of the tobacco particles), product format (loose or packaged), pH, and buffering capacity to hold the pH constant. Several lines of evidence suggest that some ingredients in popular mint-flavored products may increase nicotine absorption by increasing the permeability of the oral mucosa.[20] However, this hypothesis has not been systematically tested.

Effects of free nicotine content and pH on abuse liability

Nicotine occurs in several chemical states, including bound to hydrogen ions (mono-protonated and di-protonated) and unbound (unprotonated, unionized, and free). The amount of free nicotine is determined by total nicotine content and the pH of the product.[21] Free nicotine content affects nicotine exposure because it more readily crosses biological membranes, including oral mucosa.[22–24] In an alkaline (high pH) environment, nicotine is unbound and rapidly absorbed; in an acidic (low pH) environment, it is bound and does not cross membranes as easily. Absorption across the oral mucosa appears to be related to the amount present in the unbound "free base" form. Work by Borgerding and colleagues has shown that aqueous solutions of ST products bracket a wide pH range, from 5.0 to 8.4, with associated concentrations of unbound nicotine between <1% and 80% (Fig. 4.4).[25]

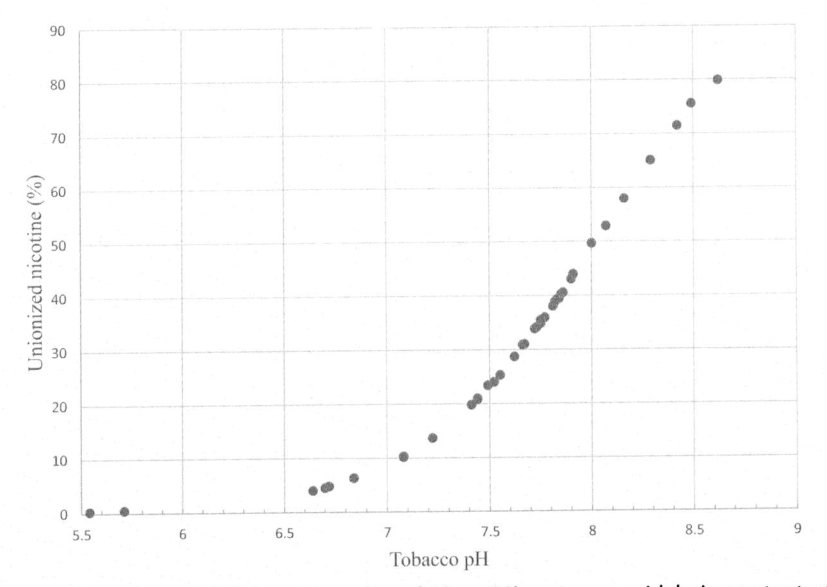

Figure 4.4 Moist smokeless tobacco brands (n = 39) may vary widely in content of rapidly absorbed, addictive unionized nicotine (500-fold range). *Based on data of Richter P, Hodge K, Stanfill S, Zhang L, Watson C. Surveillance of moist snuff: total nicotine, moisture, pH, un-ionized nicotine, and tobacco-specific nitrosamines.* Nicotine Tob Res. *2008;10(11):1645–1652.*

There is evidence that ST products deliver large amounts of nicotine, even compared to cigarettes. Benowitz and colleagues studied the extent and time course of absorption from oral snuff and chewing tobacco in comparison with smoking cigarettes and chewing nicotine gum in 10 healthy volunteers.[27] The maximum levels of nicotine were similar for ST and cigarettes (averaging 14.3 ng/mL in blood); but, because of prolonged absorption, overall nicotine exposure was twice as large after single exposures to ST than after cigarette smoking (Fig. 4.5).

Results from clinical studies conducted under well-controlled laboratory conditions suggest that free nicotine content affects ST abuse liability. For example, Fant and colleagues examined the effects of four commercially available snuff products on nicotine pharmacokinetics and pharmacodynamics in current users.[28] The products were Copenhagen, Skoal Long Cut Cherry, Skoal Original Wintergreen, and Skoal Bandits, chosen for

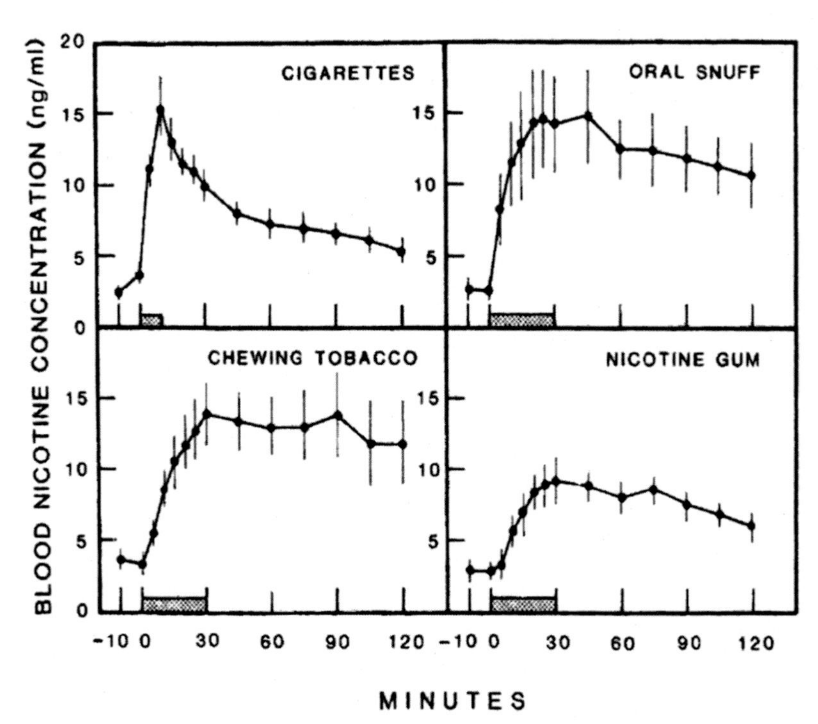

Figure 4.5 Blood nicotine concentrations during and after cigarette smoking, oral snuff, chewing tobacco, and nicotine gum (two 2 mg pieces). *Reprinted from Benowitz NL, Porchet H, Sheiner L, Jacob P, 3rd. Nicotine absorption and cardiovascular effects with smokeless tobacco use: comparison with cigarettes and nicotine gum.* Clin Pharmacol Ther. *1988;44(1):23–28.*

differences in free nicotine content. The results showed positive relations between free nicotine content and several outcomes, including plasma nicotine concentration, heart rate, and subjective effects. A limitation of this study was that the effects of free nicotine content on study outcomes were not isolated because other characteristics (flavorings, binders, sweeteners, etc.) varied between products. Nevertheless, there were orderly associations between free nicotine content and various measures of abuse liability.

Pickworth and colleagues also examined the effects of ST pH on nicotine absorption.[29] They used a referent unflavored moist snuff tobacco product obtained from a university tobacco program. All product characteristics (e.g., water content, tobacco blend, and nicotine content) were held constant across experimental products except for pH and methyl salicylate. The researchers manufactured three products: (1) ST with pH 5.4 (<0.5% free nicotine) and methyl salicylate, (2) ST with pH 7.7 (32% free nicotine) and no methyl salicylate, and (3) ST with pH 8.3 (66% free nicotine) and methyl salicylate. Participants were exposed to all three products in separate experimental sessions. Plasma nicotine concentration was higher following administration of the two products with higher pH relative to the product with lower pH. In addition, the rate of nicotine absorption was much faster in the products with higher pH levels. Notably, no differences in plasma nicotine concentration were observed between the two products with the highest pH levels. However, study limitations (e.g., small sample size, differences in methyl salicylate content between these two products) may have contributed to this finding. Nevertheless, the study found large differences in plasma nicotine in the product with the lowest pH relative to the product with the highest pH, and both had the same methyl salicylate content. The researchers concluded, "These results indicate that pH is a primary determinant of buccal nicotine absorption."

Surveillance and monitoring studies have shown that pH levels vary widely between products, resulting in percentages of free nicotine ranging from less than 1% to over 75%.[26] Such differences can affect abuse liability in various ways. For example, evidence from survey studies and industry documents suggests that "starter" products with low free nicotine are marketed to inexperienced users.[30,31] Low free nicotine may increase initiation. Evidence also suggests that consumers who begin with starter products are likely to graduate to those with higher free nicotine, which are associated with increased use and greater signs of dependence.[31] Thus, increases in the free nicotine content of ST may increase dependence.

These results indicate that nicotine content and tobacco pH are primary determinants of nicotine absorption across the oral mucosa. Evidence suggests that higher free nicotine results in larger, more rapid increases in plasma nicotine.[28,29] These effects occur independent of topography (e.g., duration of use) and other use behaviors (e.g., consuming coffee and cola) when behaviors are controlled under laboratory conditions. Additional evidence suggests that decreased free nicotine may increase the likelihood of ST initiation and increased free nicotine may increase the likelihood of dependence.[31]

Effects of format (loose vs. portioned)

The United States domestic ST, specifically moist snuff, is commercially available in two distinct forms: loose and portioned (sachets, tea bags, pouches). Users of the loose form take a pinch from the can and place it in the mouth between cheek and gum. The portioned products are usually placed, one or more at a time, also between cheek and gum. Few studies have investigated the amount of tobacco typically used, the frequency of use, and the pharmacokinetic and pharmacodynamic effects of loose versus portioned. Although there is some preliminary evidence that format affects exposure to nicotine and other tobacco toxicants,[25] a better understanding is warranted to inform public health policy. Individual users may alter their behavior when changing between loose and portioned, including frequency of use, deposition time in the mouth, amount of product per occasion, mean use per day, and total use per day. Differences in use topography of tobacco are known to affect nicotine exposure.[7,32,33]

Digard and colleagues[34] conducted a telephone survey of Swedish tobacco users and found that, among snus users, the loose form was used more frequently and in greater amounts than the portioned form, with no significant differences in duration (that is, deposition time in the mouth).[34] From daily diary information, frequency and duration were found to be more strongly correlated with nicotine and carcinogen exposure than was amount used.[33] A later study, also by Digard and colleagues[35], compared nicotine absorption following instructed use of a loose or portioned snus which was placed under the upper lip and not moved for 1 hour; nicotine absorption depended on the quantity of tobacco by weight and total nicotine content rather than format.[35] This study was limited to only three Swedish snus products (two portioned with different nicotine contents and one loose at two portion sizes) and included a very prescribed use behavior.

No prospective study has been published comparing the behavioral pharmacology (e.g., use behavior, constituent exposure) of loose to portioned format. Therefore, a clinical study was designed to increase understanding of normal use patterns and nicotine exposure (unpublished data from Hull and colleagues). Participants were asked to use their products ad libitum while several subjective and pharmacokinetic measures were taken. Differences between the two formats were observed in user topography (i.e., average amount per use, total amount used, average deposition time per use) and harmful and potentially harmful constituent exposure (e.g., nicotine absorption). Format differences were also found in product characteristics [i.e., total nicotine, free nicotine content, (4-(methylnitrosamino)-1-(3-pyridyl)-1-butanone) (NNK), nornicotine, anabasine, myosamine]. These results indicate the need for additional studies to tease out whether there is a relationship between user behaviors or product constituents of the two format types and the pharmacokinetics of nicotine.

Effects of flavors

Menthol is the most popular ST flavor in the United States.[36] It is a terpene that in vitro increases transdermal and transbuccal penetration of various compounds, including nicotine and the carcinogens N-nitrosonornicotine (NNN), and NNK.[37–42] Terpenes facilitate the uptake of both lipophilic and hydrophilic compounds across the oral mucosa by various mechanisms.[42] Squier and colleagues investigated the effects of menthol on the penetration of nicotine and NNN in porcine oral mucosa.[40] They found that when menthol was added at a concentration of 0.08% to two solutions containing either 3% nicotine or 0.01% NNN, it significantly increased penetration of both compounds. These findings are particularly noteworthy, given that the concentrations of all three were comparable to those typically found in tobacco products. Further, the durations of menthol exposure (30 minutes, 1 hour, and 2 hours) were much more comparable to those in tobacco use than to those in most in vitro permeation studies (e.g., 6–12 hours). Further clinical studies are needed to investigate the effects of menthol on membrane permeability and user exposure to nicotine and other constituents.

Future of clinical laboratory studies

The absorption of nicotine from ST products is contingent upon product characteristics and user behavior. Although evidence suggests that certain characteristics (e.g., pH, menthol flavoring) may affect absorption across the oral mucosa, clinical research is needed to confirm this potential and determine the extent of impact. User-specific modulators also need to be investigated, particularly how factors such as experience level (e.g., novice youth vs. established users, exclusive ST users vs. dual- or polytobacco users) affect absorption in the diverse user population. Although the ST user population is small compared to the total tobacco user population, it is not negligible. There is still a lack of understanding about how ST products work in humans on a pharmacological level. Moreover, while there is a downward trend in smoking (likely a combination of cessation and fewer new users), there is no such trend in ST users, suggesting that more research is needed on mechanisms of dependence. This is especially important in light of proposed regulation in the United States to dramatically reduce smoking rates by lowering the nicotine in combustible cigarettes to minimally or nonaddictive levels, which could result in smokers switching to ST. The need for research seems particularly urgent in this area, as ST product manufacturers have submitted applications for modified risk claims to the Food and Drug Administration, indicating that they might be interested in increasing market share. It is possible that the tobacco industry wants the public to view ST products as good candidates for switching in a market where high-nicotine cigarettes could no longer be available.

References

1. Benowitz NL. Nicotine addiction. *N Engl J Med*. 2010;362(24):2295–2303.
2. Henningfield JE, Fant RV, Tomar SL. Smokeless tobacco: an addicting drug. *Adv Dent Res*. 1997;11(3):330–335.
3. Stolerman IP, Jarvis MJ. The scientific case that nicotine is addictive. *Psychopharmacology (Berl)*. 1995;117(1):2–10.
4. US Department of Health and Human Services. The health consequences of smoking: nicotine addiction. In: *A Report of the Surgeon General*. 1988.
5. Henningfield JE, Miyasato K, Jasinski DR. Abuse liability and pharmacodynamic characteristics of intravenous and inhaled nicotine. *J Pharmacol Exp Ther*. 1985;234(1):1–12.
6. Perkins KA, Grobe JE, Epstein LH, Caggiula A, Stiller RL, Jacob RG. Chronic and acute tolerance to subjective effects of nicotine. *Pharmacol Biochem Behav*. 1993;45(2):375–381.
7. Hatsukami DK, Keenan RM, Anton DJ. Topographical features of smokeless tobacco use. *Psychopharmacology (Berl)*. 1988;96(3):428–429.
8. Henningfield JE, Radzius A, Cooper TM, Clayton RR. Drinking coffee and carbonated beverages blocks absorption of nicotine from nicotine polacrilex gum. *J Am Med Assoc*. 1990;264(12):1560–1564.
9. Digard H, Proctor C, Kulasekaran A, Malmqvist U, Richter A. Determination of nicotine absorption from multiple tobacco products and nicotine gum. *Nicotine Tob Res*. 2012;15(1):255–261.
10. Henningfield JE, Goldberg SR. Nicotine as a reinforcer in human subjects and laboratory animals. *Pharmacol Biochem Behav*. 1983;19(6):989–992.

11. Henningfield JE, Higgins ST. The influence of behavior analysis on the surgeon general's report, the health consequences of smoking: nicotine addiction. *Behav Anal.* 1989; 12(1):99–101.
12. Henningfield JE, Nemeth-Coslett R. Nicotine dependence. Interface between tobacco and tobacco-related disease. *Chest.* 1988;93(2 Suppl):37s–55s.
13. CDC. Smokeless. *Tobacco Use in the United States* [Fact Sheet]; 2019. https://www.cdc.gov/tobacco/data_statistics/fact_sheets/smokeless/use_us/index.htm.
14. Piper ME, Piasecki TM, Federman EB, et al. A multiple motives approach to tobacco dependence: the Wisconsin Inventory of Smoking Dependence Motives (WISDM-68). *J Consult Clin Psychol.* 2004;72(2):139–154.
15. Shiffman S, Waters A, Hickcox M. The nicotine dependence syndrome scale: a multidimensional measure of nicotine dependence. *Nicotine Tob Res.* 2004;6(2):327–348.
16. Heatherton TF, Kozlowski LT, Frecker RC, Rickert W, Robinson J. Measuring the heaviness of smoking: using self-reported time to the first cigarette of the day and number of cigarettes smoked per day. *Br J Addict.* 1989;84(7):791–799.
17. Liu ST, Nemeth JM, Klein EG, Ferketich AK, Kwan MP, Wewers ME. Adolescent and adult perceptions of traditional and novel smokeless tobacco products and packaging in rural Ohio. *Tob Control.* 2014;23(3):209–214.
18. Hatsukami DK, Zhang Y, O'Connor RJ, Severson HH. Subjective responses to oral tobacco products: scale validation. *Nicotine Tob Res.* 2013;15(7):1259–1264.
19. Cappelleri JC, Bushmakin AG, Baker CL, Merikle E, Olufade AO, Gilbert DG. Confirmatory factor analyses and reliability of the modified cigarette evaluation questionnaire. *Addict Behav.* 2007;32(5):912–923.
20. Chen C, Isabelle LM, Pickworth WB, Pankow JF. Levels of mint and wintergreen flavorants: smokeless tobacco products vs. confectionery products. *Food Chem Toxicol.* 2010;48(2):755–763.
21. Henningfield JE, Radzius A, Cone EJ. Estimation of available nicotine content of six smokeless tobacco products. *Tob Control.* 1995;4(1):57–61.
22. Chen LH, Chetty DJ, Chien YW. A mechanistic analysis to characterize oramucosal permeation properties. *Int J Pharm.* 1999;184(1):63–72.
23. Nielsen HM, Rassing MR. Nicotine permeability across the buccal TR146 cell culture model and porcine buccal mucosa in vitro: effect of pH and concentration. *Eur J Pharm Sci.* 2002;16(3):151–157.
24. Nair MK, Chetty DJ, Ho H, Chien YW. Biomembrane permeation of nicotine: mechanistic studies with porcine mucosae and skin. *J Pharm Sci.* 1997;86(2):257–262.
25. Borgerding MF, Bodnar JA, Curtin GM, Swauger JE. The chemical composition of smokeless tobacco: a survey of products sold in the United States in 2006 and 2007. *Regul Toxicol Pharmacol.* 2012;64(3):367–387.
26. Richter P, Hodge K, Stanfill S, Zhang L, Watson C. Surveillance of moist snuff: total nicotine, moisture, pH, un-ionized nicotine, and tobacco-specific nitrosamines. *Nicotine Tob Res.* 2008;10(11):1645–1652.
27. Benowitz NL, Porchet H, Sheiner L, Jacob 3rd P. Nicotine absorption and cardiovascular effects with smokeless tobacco use: comparison with cigarettes and nicotine gum. *Clin Pharmacol Ther.* 1988;44(1):23–28.
28. Fant RV, Henningfield JE, Nelson RA, Pickworth WB. Pharmacokinetics and pharmacodynamics of moist snuff in humans. *Tob Control.* 1999;8(4):387–392.
29. Pickworth WB, Rosenberry ZR, Gold W, Koszowski B. Nicotine absorption from smokeless tobacco modified to adjust pH. *J Addict Res Ther.* 2014;5(3):1000184.
30. Connolly GN. The marketing of nicotine addiction by one oral snuff manufacturer. *Tob Control.* 1995;4(1):73.
31. Tomar SL, Giovino GA, Eriksen MP. Smokeless tobacco brand preference and brand switching among US adolescents and young adults. *Tob Control.* 1995;4(1):67.

32. Hatsukami DK, Anton D, Callies A, Keenan R. Situational factors and patterns associated with smokeless tobacco use. *J Behav Med*. 1991;14(4):383–396.
33. Lemmonds CA, Hecht SS, Jensen JA, et al. Smokeless tobacco topography and toxin exposure. *Nicotine Tob Res*. 2005;7(3):469–474.
34. Digard H, Errington G, Richter A, McAdam K. Patterns and behaviors of snus consumption in Sweden. *Nicotine Tob Res*. 2009;11(10):1175–1181.
35. Digard H, Proctor C, Kulasekaran A, Malmqvist U, Richter A. Determination of nicotine absorption from multiple tobacco products and nicotine gum. *Nicotine Tob Res*. 2013;15(1):255–261.
36. Bonhomme MG, Holder-Hayes E, Ambrose BK, et al. Flavoured non-cigarette tobacco product use among US adults: 2013–2014. *Tob Control*. 2016;25(Suppl 2):ii4–ii13.
37. Kunta JR, Goskonda VR, Brotherton HO, Khan MA, Reddy IK. Effect of menthol and related terpenes on the percutaneous absorption of propranolol across excised hairless mouse skin. *J Pharm Sci*. 1997;86(12):1369–1373.
38. Song YH, Gwak HS, Chun IK. The effects of terpenes on the permeation of lidocaine and ofloxacin from moisture-activated patches. *Drug Deliv*. 2009;16(2):75–81.
39. Shojaei AH, Khan MA, Lim G, Khosravan R. Transbuccal permeation of a nucleoside analog, dideoxycytidine: effects of menthol as a permeation enhancer. *Int J Pharm*. 1999;192(2):139–146.
40. Squier CA, Mantz MJ, Wertz PW. Effect of menthol on the penetration of tobacco carcinogens and nicotine across porcine oral mucosa ex vivo. *Nicotine Tob Res*. 2010;12(7):763–767.
41. Azzi C, Zhang J, Purdon CH, et al. Permeation and reservoir formation of 4-(methylnitrosamino)-1-(3-pyridyl)-1-butanone (NNK) and benzo [a] pyrene (B [a] P) across porcine esophageal tissue in the presence of ethanol and menthol. *Carcinogenesis*. 2005;27(1):137–145.
42. Hassan N, Ahad A, Ali M, Ali J. Chemical permeation enhancers for transbuccal drug delivery. *Expert Opin Drug Deliv*. 2010;7(1):97–112.

CHAPTER FIVE

Smokeless tobacco treatment: measures, interventions, recommendations, and future directions

Herbert H. Severson[1], Judith S. Gordon[2], Laura Akers[1], Devon Noonan[3]

[1]Oregon Research Institute, Eugene, OR, United States
[2]University of Arizona, Tucson, AZ, United States
[3]Duke University, Durham, NC, United States

Introduction

While cigarette smoking is far more prevalent than smokeless tobacco (ST) use in the United States, there are almost 9 million regular ST users, and the rate has remained stable or increased over the last 10 years.[1] Achieving abstinence is the most effective way to prevent morbidity and mortality related to ST use.[2] Most published clinical trials evaluating interventions for ST use have occurred in the United States, making it difficult to generalize the findings to other nations with different types and patterns of use. While ST is prevalent in other parts of the world, the differences in products, patterns of use, and regulations on manufacture and marketing are beyond the scope of this chapter. Therefore, the focus is on cessation studies conducted in the United States. It deals only with adults (age 18 or older) and does not review the literature on cessation in adolescents or children. The Surgeon General's report *Preventing Tobacco Use Among Youth and Young Adults*[3] and *Smokeless Tobacco and Public Health: A Global Perspective*[4] offer excellent reviews of preventive and cessation interventions for youth.

Behavioral and pharmacologic interventions to treat ST use have been evaluated in multiple studies.[5] Outcomes have been measured by short- and long-term (≥6 months) tobacco abstinence rates and by self-reported levels of tobacco craving and nicotine withdrawal.[5] This chapter describes unique assessment issues for ST cessation. These include measurement of use, addiction and dependence, readiness or motivation to quit, and

Smokeless Tobacco Products
ISBN: 978-0-12-818158-4
https://doi.org/10.1016/B978-0-12-818158-4.00005-4

outcome such as cessation, craving, and withdrawal. Emphasis is on practical suggestions for cessation programs and key features of efficacious therapies reported in peer-reviewed journals.

Recommendations are based on the extant literature and our own clinical experience. Recommendations for measures, procedures, target outcomes, and key features of interventions provide a framework for conducting research to assist ST users in quitting. A broad array of resources and interventions that can be adapted to the needs of ST users and healthcare professionals are offered. Finally, we present a compilation of future directions for research and practice. Our goal is to suggest directions that need further exploration and evaluation as we seek to improve the outcomes of ST cessation interventions.

Challenges in measuring smokeless tobacco use and dependence

Researchers who conduct studies on smoking have advantages in measuring use over those studying ST. Measuring cigarette use is much easier than ST due to the standards for cigarettes and their packaging. Cigarettes vary little in size and constituents, and generally are sold in packs of the same quantity. Therefore, it is fairly straightforward to quantify and collect use data from smokers (e.g., number of cigarettes or packs smoked in some time interval). Assessment of ST use is more difficult because of the variability of products and packaging. Chewing tobacco is sold in foil pouches, and snuff and snus are sold in tins. Quantifying use is challenging and fraught with variability. Researchers must rely on users' reports of number of chews or dips per day, but the amount of tobacco contained in a chew or dip may vary considerably. For these reasons, the most common method is to ask how long a tin or pouch lasts, or how many tins or pouches are used per week, although some studies also ask the user to report on the number of chews or dips per day.

Other measures have been adapted to assess the readiness of ST users to quit. Severson and colleagues have been using an adaption of the Biener and Abrams' Contemplation Ladder,[6] which provides a simple method for the user to mark their level of readiness to quit on a 11-point scale of 0—10, with 0 indicating not ready and 10 indicating ready now. We have found that the level of readiness has been a modest predictor of actual quitting in several of our studies evaluating cessation interventions.[7,8] Levels of confidence in quitting can also be done by an ordinal scale of 0—10. Chewers and dippers are asked how confident they are that they will not be using ST a year from now. Both the readiness and confidence in quitting scales can provide ongoing measures for comparison between baseline and follow-up assessments. Scores of 0–7 are categorized as low to moderate and 8 to 10 as high.[9]

Assessing dependence on ST has been done largely by the adaption of measures designed for smokers. The most widely employed measure of tobacco dependence in clinical use and research remains the Fagerström Tolerance Questionnaire (FTQ).[10] It measures physiologic and behavioral parameters. Items include cigarettes per day, time to first cigarette in the morning, frequency of inhalation, and questions related to dependence behaviors such as difficulty in refraining from use. The FTQ has relatively low internal consistency and reliability, but the item measuring time to first cigarette correlates well with other scales and the entire FTQ.

The first scales to assess ST dependence were modified versions of the FTQ[11] (Table 5.1). The Fagerström Tolerance Questionnaire for Smokeless Tobacco (FTQ-ST) scores are positively correlated with serum cotinine and amount of tobacco used. However, poor concordance was found between the FTQ-ST and the diagnosis of nicotine dependence as measured by the Diagnostic Interview Schedule (DIS), a structured interview based on the American Psychiatric Association *Diagnostic and Statistical Manual of Mental Disorders*, 4th edition (DSM-IV) criteria.[12] More recent ST dependence scales have items drawn from the literature on smoking motives, behavioral patterns, and withdrawal symptoms and expectations.[13] Severson and colleagues created the Severson Smokeless Tobacco Dependency Scale (SSTDS), which includes items from the FTQ, items assessing behavioral patterns of ST use, and anticipated withdrawal symptoms (Table 5.2). The Glover–Nilsson Smoking Behavioral Questionnaire (GN-SBQ) focuses on behavioral patterns such as "the rituals that surround smoking, and the relationship that the smoker develops with the cigarette."[14] Glover and colleagues created a version for ST users which is intended to be used with a measure of physical dependence scale, the Glover–Nilsson Smokeless Tobacco Behavioral Questionnaire (GN-STBQ). A recent study compared the FTQ-ST, the SSTDS, and the GN-STBQ and concluded that the FTQ-ST appears to best measure the construct of physical dependence and was correlated with serum nicotine and cotinine concentrations.[13] The SSTDS was the only scale that was independently associated with withdrawal and craving.

Table 5.1 Fagerström tolerance questionnaire for smokeless tobacco (FTQ-ST).

Item	Response	Points
1. After a normal sleeping period, do you use smokeless tobacco within 30 minutes of waking?	Yes	1
	No	0
2. Is it difficult for you not to use smokeless where its use would be unsuitable or restricted?	Yes	1
	No	0
3. Do you use smokeless when you are sick or have mouth sores?	Yes	1
	No	0
4. Nicotine content	Low	1
	Medium	2
	High	3
5. How many days does a tin/can last you?	6 to 7	1
	3 to 5	2
	<3	3
6. On average how many minutes do you keep a fresh dip or chew in your mouth?	10 to 19	1
	20 to 30	2
	>30	3
7. How often do you swallow tobacco juices?	Never	0
	Sometimes	1
	Always	2
8. Do you keep a dip or chew in your mouth almost all the time?	Yes	1
	No	0
9. Do you experience strong cravings for a dip/chew when you go for more than 2 hours without one?	Yes	1
	No	0
10. On average how many dips or chews do you take each day?	1 to 9	1
	10 to 15	2
	>15	3

The total score was calculated as the sum of the 10 items. Possible scores ranged from 4 to 19. For the current sample, the Cronbach coefficient alpha was 0.60.
Adapted from Fagerström KO, Schneider NG. Measuring nicotine dependence: a review of the Fagerström Tolerance Questionnaire. *J Behav Med.* 1989;12(2):159–182.

Since that comparative study was published, new measures have been developed. The Fagerström Test for Nicotine Dependence-Smokeless Tobacco (FTND-ST) adapted from the Fagerström Test for Nicotine Dependence contains 6 items. It was tested with 42 ST users and was found to have significant correlations with serum cotinine levels; however, the internal

Table 5.2 Severson smokeless tobacco dependence scale (SSTDS) [N = 223].

Item	Response	Points	Observed No. (%)
1. How many days does a tin or pouch last you?	<1	NA	15 (7)
	1		73 (33)
	2		61 (27)
	3		27 (12)
	4		8 (4)
	5		10 (4)
	6		4 (2)
	7		21 (9)
	>7		4 (2)
2. Do you experience strong cravings for a dip/chew when you go more than 2 hours without one?	Yes	1	188 (84)
	No	0	35 (16)
3. How soon after you wake up do you use chew/snuff?	0–30 min	1	159 (71)
	>30 min	0	64 (29)
4. When you go without a dip or chew do you find yourself getting anxious more quickly?	Never	0	13 (6)
	Seldom	1	20 (9)
	Sometimes	2	65 (29)
	Often	3	81 (36)
	Always	4	44 (20)
5. When you go without a dip or chew do you find yourself getting drowsy more quickly?	Never	0	49 (22)
	Seldom	1	78 (35)
	Sometimes	2	68 (30)
	Often	3	22 (10)
	Always	4	6 (3)
6. I use more snuff/chew when I am worried about something.	Not at all	0	16 (7)
	A little	1	58 (26)
	Quite a bit	2	86 (39)
	Very much so	3	63 (28)
7. I use more snuff/chew when I am rushed and have lots to do.	Not at all	0	16 (7)
	A little	1	67 (30)
	Quite a bit	2	67 (30)
	Very much so	3	73 (33)
8. I get a definite lift and feel more alert when using snuff/chew.	Not at all	0	20 (9)
	A little	1	73 (33)
	Quite a bit	2	79 (35)
	Very much so	3	51 (23)

The total score was calculated as the sum of items 2 through 8. Possible scores ranged from 0 to 19. For the current sample, the Cronbach coefficient alpha was 0.69.

consistency was found to be low.[15] The Oklahoma Scale for Smokeless Tobacco Dependence (OSSTD, Table 5.3), adapted from the Wisconsin Inventory of Smoking Dependence Motives (WISDM-68),[16] contains 23 items and 7 latent constructs. It was tested on 100 adult male ST users and significant correlations were reported between total score, cotinine levels, and ST use.[17] The internal consistency ($\alpha = 0.925$) indicated better reliability than FTND-ST. All subscales except affective enhancement were positively associated with the FTND-ST. Mushtaq and colleagues also explored the utility of three ST use indices as dependence measures.[18] Modified from the Heaviness of Smoking Index, time to first chew or dip of the day (TTFD), number of cans used per week (CPW), and dips/chews per day (DPD) were used to create three use indices: the heaviness of ST use index (HSTI), ST dependence index (STDI), and ST quantity frequency index (ST-QFI). The HSTI and the STDI were highly correlated with the FTND-ST, and ST use indices were predictors of cotinine concentration. Finally, the Tobacco Dependence Screener (TDS) was adapted and tested in ST users.[19] It showed strong reliability and concurrent validity compared to the FTND-ST. Significant correlations were found between salivary cotinine levels and the TDS for ST users. Given the small sample sizes of all these studies (100 or less), resulting in only moderate power, they should be replicated with larger samples.

The criteria for measuring abstinence have also been adapted from smoking cessation research. The gold standards are 7-day point prevalence, 30-day prolonged abstinence, and sustained abstinence at follow-up. Essentially, the measures used for ST cessation mirror those for smoking cessation.[20,21] For point prevalence and 30-day prolonged abstinence of ST use, studies use questions that ask the participant to report if they have had any chew or snuff in the past 7 or the past 30 days.[7,22–24] Sustained abstinence is measured by self-reported consecutive point prevalence (e.g., no tobacco use for the past 7 days at both the 3-month and 6-month assessments).[7] Most assessments for ST cessation also ask about smoking behaviors. Most studies define abstinence as report of no tobacco use, including ST and smoking.[25] This is important, as a modest percent of ST users (5%–10%) also report smoking.[26]

Interventions

Research on ST cessation

While significantly fewer studies have evaluated ST cessation than smoking cessation, numerous studies have examined diverse approaches to

Table 5.3 Oklahoma scale for smokeless tobacco dependence (OSSTD): Subscales and items.

	Item No.[a]	Subscales and items
Positive dependence motives	Loss of control/craving	
	1	Chew/dip controls me.
	12	I'm really hooked on chew/dip.
	4	It is hard to ignore an urge to chew/dip.
	10	I frequently crave chew/dip.
	Tolerance/automaticity	
	16	Other chewers/dippers would consider me a heavy chewer/dipper.
	18	I chew/dip within the first 30 min of awakening in the morning.
	13	I find myself reaching for chew/dip without thinking about it.
	19	Sometimes I am not aware that I am chewing/dipping.
Secondary dependent motives	Affective enhancement	
	2	Chewing/dipping improves my mood.
	21	Chewing/dipping really helps me feel better if I have been feeling down.
	22	Chewing/dipping makes me feel good.
	Affiliative attachment	
	7	Chew/dip keeps me company, like a close friend.
	15	I would feel alone without my chew/dip.
	3	Very few things give me pleasure each day like chewing/dipping.
	Cognitive enhancement	
	5	I chew/dip when I really need to concentrate.
	9	Chewing/dipping helps me stay focused.
	20	Chewing/dipping helps me think better.

(Continued)

Table 5.3 Oklahoma scale for smokeless tobacco dependence (OSSTD): Subscales and items.—cont'd

Item No.[a]	Subscales and items
Weight control	
6	I rely upon chewing/dipping to control my hunger and eating.
11	Weight control is a major reason that I chew/dip.
23	Chewing/dipping keeps me from over eating.
Cue exposure	
8	There are particular sights and smells that trigger strong urges to chew/dip.
14	I crave chew/dip at certain times of the day.
17	Some things are very hard to do without chewing/dipping.

[a]Item numbers from the original WISDM-68.

helping chewers quit. We reviewed multiple intervention studies and found that they fell into three categories: (1) those delivered in dental settings, (2) those delivered in military settings, and (3) self-help interventions delivered by a variety of methods, including behavioral and/or pharmacotherapy. Below we discuss these three categories and the efficacy of oral substitutes and pharmacological aids.

Unique aspects of ST cessation and opportunities for intervention

While the interventions for ST users are largely based on successful methods for smokers, there are unique aspects of using ST that warrant attention. Regular use of ST products, particularly moist snuff, often results in the development of oral lesions at the site where the tobacco is held in the mouth. These are called leukoplakic lesions. Research has shown that the prevalence and severity is directly correlated with years of use and level of daily use.[27] One study in dental clinics found that 78% of daily users of snuff had identifiable lesions at the site where the product was held.[28] The cause of the lesions has been attributed to the high levels of tobacco-specific nitrosamines (TSNA), the prime carcinogens in ST.[29] Since tobacco companies have reduced the levels of TSNA and other nitrosamines in their products, and snus is marketed as a product with much lower levels of

TSNA, the relationship between regular use and development of oral lesions may be changing. However, the high likelihood of a regular user having an oral lesion provides a unique opportunity to relate the person's use of these products to a visible health effect. This "teachable moment" has been a key element in dental office interventions, as visual feedback about a lesion can motivate the user to quit. The brief screening and identification of oral lesions has been a key element in the dental interventions described below.

Another unique element of interventions for ST users is the need to identify oral substitutes to use during the cessation process. In addition to the craving for nicotine that is common across all withdrawal from regular tobacco use, the ST user often misses having the tobacco in their mouth and asks for oral substitutes. Relative to smoking cigarettes, there are discrete sensory stimulations and cures associated with ST use, and some authors have noted product-specific behaviors and stimuli as a function of dependence.[30] Our research team has suggested a wide range of options for nontobacco substitutes, including ground up mint leaves and other things that are sold in tins to resemble snuff and are specifically marketed for this purpose.[31] Research on substitutes is covered later in this section. Some users find that sugarless gum and candies, cinnamon sticks, sunflower seeds, and toothpicks also suffice.

In addition, ST users may develop different patterns of use and dependence than smokers. While smokers have multiple restrictions on when and where they can smoke, ST users can use these products virtually 24 hours per day. This near-constant exposure can result in higher levels of dependence and behavioral associations between ST use and many daily activities. These combined physiological and behavioral factors present additional challenges, suggesting that a combination of pharmacotherapy, oral substitutes, and behavioral coaching may be the most effective approach.[25,32,33]

Finally, most ST users are male and more likely to be rural and work in certain professions (e.g., ranching, military, baseball).[25,34] Therefore, interventions may need to employ different types of behavioral approaches or different levels of pharmacotherapy. The demographics of ST users also provide opportunities to deliver interventions in specific settings such as among military personnel and baseball players.[25] Research on these special populations is described below.

Intervention in dental settings

Dental offices offer a unique opportunity to provide brief counseling and support for quitting. Organized dentistry has long recognized the opportunity for oral health professionals to address tobacco use,[35–38] and the American Dental Association urges dentists to assist patients with cessation.[35] The dental team routinely provides educational and preventive services to patients; this model can be extended to tobacco-related topics. The dental team can relate oral health and systemic problems to tobacco use and provide evidence-based brief interventions.

These interventions are typically based on the "5 A's" advocated by the Clinical Practice Guideline *Treating Tobacco Use and Dependence*,[39] an adaptation from use in medical settings.[40] The 5 A's are Ask (systematically identify tobacco users at every visit), Advise (urge all tobacco users to quit), Assess (determine willingness to make an attempt to quit), Assist (aid the patient in quitting, which might involve counseling, self-help materials, and medication), and Arrange (ensure follow-up contact). The AAR model uses the same definitions of Ask and Advise, but replaces Assess, Assist, and Arrange with some type of Referral to a cessation program. Referrals have ranged from "passive" suggestions to use available cessation programs (ranging from local group treatment to tobacco quitline) to "proactive" referrals to tobacco quitlines, web-based or text-based programs in which the provider faxes a referral or electronically enrolls the user into the program.

Dental hygiene visits are a unique opportunity to assess for tobacco use and offer cessation advice and support. Dental hygienists have been receptive to training and to providing interventions, and research shows that they can be effective at helping ST users to quit.[36,41,42] The length of the hygiene visit and the focus on health education are conducive to intervention.

Dental setting interventions in both military and civilian populations have been effective in increasing tobacco abstinence rates among ST users. In a study of 24 US military dental clinics, 785 users were randomized to usual care or telephone counseling by a trained cessation counselor. They received assistance in quitting (if desired) along with a mailed videotape and a military-tailored self-help guide.[8] For the phone counseling group, the first call occurred about 1 week after the dental visit. Individuals accepting materials were offered two or more calls coinciding with receipt of the mailed materials and their quit date. Subjects in this program were significantly more likely to be abstinent from all tobacco as assessed by repeated point prevalence at both 3 and 6 months (25.0%) and were significantly more likely to be abstinent from ST for 6 months as assessed by prolonged abstinence (16.8%), compared with usual care (7.6% and 6.4%, respectively). Identifying active duty ST users during preventive health screenings and providing an ST treatment manual, a video, and several supportive phone calls from a cessation counselor has been shown to increase tobacco abstinence rates compared to usual care.[8] Usual care involved being counseled by the dental provider to quit use of ST and provision of a referral to the local military installation tobacco cessation program. In this study, the 3- and 6-month tobacco abstinence rates in the intervention group were double those in usual care (41% vs. 17% at 3 months) but not significantly different at 6 months (37% vs. 19%).[8]

In another study conducted at 11 dental clinics, 518 ST users were randomized to usual care or a behavioral intervention incorporating an oral examination with feedback, advice to quit from a hygienist and dentist, a self-help manual, a video, setting a quit date, telephone support from a counselor, a free helpline, and six newsletters.[36] The intervention significantly increased long-term abstinence rates at both 3 and 12 months (12.5% and 18.4%) compared to usual care. The authors cite the feedback from oral exams as a key motivational factor for getting patients to try quitting.[36]

In a study of 68 private dental clinics, 2160 tobacco users (ST users, smokers, and dual-users) were randomized to one of three conditions: 5 A's, AAR, or usual care.[43] Providers (dentists and dental hygienists) in the 5 A's clinics Asked all patients about their tobacco use; Advised tobacco users to quit; Assessed tobacco users' readiness to quit; Assisted those ready to quit by developing a quit plan, providing written self-help materials, and discussing pharmacotherapy; and Arranged a proactive referral to a tobacco quitline. Providers in the AAR condition completed the same Ask and Advise steps as in the 5 A's condition, but then only Referred (proactively) tobacco users interested in quitting to a quitline. Usual care clinics provided standard care. Results showed that participants in the two intervention conditions combined were more likely to report cessation of tobacco use, as measured by 7-day point prevalence at 12-month follow-up (12% vs. 8%, $P<.01$) and 9-month prolonged abstinence (3% vs. 2, $P<.10$). The results were inconclusive regarding the effectiveness of 5 A's versus AAR.

The results from these studies indicate that the dental office team can be effective at helping ST users to quit. However, adoption of evidence-based interventions in dental settings has lagged. Two recent articles reported that only a small fraction of tobacco users received advice to quit from a dental provider in 2010.[44,45] Of a sample of 5147 tobacco users from the National Health Interview Survey, Danesh and colleagues reported that only 11.8% had received advice to quit from their dental provider, compared to 50.7% from a physician.[44] Agaku and colleagues analyzed data from the Tobacco

Use Supplement to the Current Population Survey. In the sample of more than 35,000 tobacco users, only 31.2% had received advice to quit from a dentist (vs. 64.8% from a physician).[45] Of those who did receive advice, only 24.5% received any type of assistance: 13.8% were referred to a quitline; 10.0% were referred to a cessation class, a program, or counseling; 12.4% received a prescription or recommendation for pharmacotherapy; and only 9.8% were asked to set a quit date. Thus, there is much room for improvement in disseminating, implementing, and maintaining evidence-based tobacco cessation interventions in dental settings. Further, as reimbursement and insurance for cessation has increased exponentially in medical settings, most dental settings have been excluded from the expansion.[46] Given the cost-effectiveness and relatively low cost of covering tobacco dependency treatments in dental settings, efforts should be undertaken to mandate this coverage.[47] Increased reimbursement for cessation counseling in dental practices may be a viable way to increase cessation services given there.

Interventions for active duty military personnel

Tobacco use rates in the US military are higher than in the rest of the US population.[48,49] Effective interventions focusing on the treatment of ST dependence are critical for reducing adverse health consequences among military personnel. In a study of recruits entering basic military training, which disallows tobacco, 33,215 subjects were randomized to either a tobacco use intervention including an ST component or to a health education control.[50] The ST component included a discussion of the positive changes since quitting upon entering basic training, information about the negative consequences of ST use, a visual demonstration of negative health effects from regular use of chew and snuff, encouragement to use oral substitutes (nonnicotine and nontobacco herbal chew), and discussion of the progression from ST to other tobacco products. ST users in the intervention group were significantly more likely than ST users in the control group to be continuously abstinent at follow-up.

Cigrang and colleagues conducted a pilot study of a population-based health intervention to help military personnel quit their use of ST.[51] Sixty active-duty male participants were identified as users during their annual preventive health screening and randomly assigned to minimal-contact intervention or usual care. Intervention participants were proactively contacted by phone and recruited, using a motivational interviewing style,[52] for a cessation program consisting of a treatment manual, video, and two supportive phone calls from a cessation counselor. Sixty-five percent agreed to participate in the intervention. Three- and 6-month follow-up contacts found that the cessation rates reported by intervention participants were double those reported by participants receiving usual care (41% vs. 17% at 3 months, 37% vs. 19% at 6 months). This pilot study supported the use of proactive recruitment with a motivational interviewing approach to offer treatment in military settings.

A full-scale evaluation of ST cessation in military personnel was reported by Severson and colleagues.[8] Participants were recruited from 24 military dental clinics across the United States during the annual dental examination.[8] The study is described earlier in the Interventions in Dental Settings section, but it is important to note that the materials provided to participants were specifically created for this population of ST users. The results demonstrated that brief phone counseling following guidelines of motivational interviewing[53] supplemented with tailored video and print materials can have a significant treatment effect in personnel from all branches of military service. The behavioral treatment included an ST cessation manual,[54] a videotape cessation guide tailored for military personnel,[55] and three 15-minute telephone counseling sessions using motivational interviewing methods. The results clearly indicate that a minimal-contact behavioral treatment can significantly reduce ST use in military personnel. The study also showed the dental visit as a prime recruitment opportunity for intervention.

Interventions for baseball players

The use of ST has long been associated with baseball in the United States. Players have been cited as having very high levels of use[56,57] ranging from 24% to 39% depending on year of survey and league. This high rate or use makes them a target for interventions, and several studies have evaluated interventions to help college and professional players quit.

Walsh and colleagues conducted a study to determine the efficacy of a college-based intervention targeting athletes.[58] Sixteen colleges were matched for prevalence of ST use in their combined football and baseball teams, and within-college pairs were randomly assigned to an intervention or control condition. The intervention was a team-based program done in the college athletic facilities, where a dentist performed an oral soft tissue exam of each member, advised users to quit, and pointed out ST-related tissue changes in the player's own mouth. Players were also shown photographs of oral lesions and facial disfigurement from oral cancer. They were given a self-help guide for quitting and offered a single individual counseling session by the dental hygienist. Players interested in quitting were also offered 2 mg nicotine gum to mitigate withdrawal symptoms. The self-reported cessation results were 35% for intervention colleges and 16% for control colleges by an intent-to-treat analysis.[58] The authors concluded that the intervention was effective, especially for those who reported higher levels of use.[58]

Another study evaluated the efficacy of an athletic trainer-directed intervention.[59] This was less successful in getting players to quit. Athletes at 52 California colleges were stratified and cluster randomized to the intervention or control condition. Intervention consisted of several components aimed at both trainers and athletes. The trainers participated in a video conference and received follow-up newsletters and training manuals containing information on the adverse effects of ST use, how to motivate the team to quit, and effective cessation treatments. An additional dental component involved using the mandatory health screening as an opportunity to advise each player to stop using tobacco. ST users received oral exams to give them feedback on possible effects of use and offered individual sessions of cessation counseling. Another unique feature was having a student athlete peer lead a 1-hour educational team meeting using video and slides to facilitate a discussion about making a choice not to use tobacco. The intervention resulted in some reduction of ST initiation and self-reported levels of use but did not show a significant effect on cessation.[59]

A program for promoting cessation among professional baseball players was initiated during spring training and funded by the Robert Wood Johnson Foundation for several years. Called the National Spit Tobacco Education Program (NSTEP), it developed quitting materials and was largely facilitated by the athletic trainers for each professional team. A pilot study for implementation is described by Greene and colleagues,[60] but there was no evaluation. The program involved having well-known, retired baseball players, some of whom had suffered from oral cancer, speak to the players in the clubhouse at spring training. This was followed by feedback from oral exams that pointed out oral lesions to users. Anecdotal information suggested that the program motivated many players to quit, but the challenges of working in the framework of professional baseball and the voluntary nature of the intervention did not allow for evaluation of efficacy. There remains a need to evaluate interventions for this special population.

Self-help interventions

Manual versus telephone-based

Telephone support from trained counselors, along with self-help materials, can enhance tobacco abstinence rates among adult ST users. In a study randomizing 1069 ST users to a self-help manual-only condition or assisted self-help (ASH), the ASH intervention resulted in significantly higher ST quit rates (23.4% vs. 18.4%) and all tobacco quit rates (21.1% vs. 16.5%) at 6 months using an intent-to-treat model.[61] The ASH condition included an ST cessation manual, a video, and two support phone calls. The phone call support by counselors was a key ingredient for improving success in quitting. This study demonstrated that low-cost self-help materials followed up with telephone support provide a scalable intervention that can reach a sizable proportion of ST users. In a randomized control trial (RCT) of a phone-based intervention, 406 adult ST users in the US Midwest were randomized to a tobacco quitline with self-help combined with proactive phone counseling emphasizing support, problem-solving, and use of cognitive-behavioral strategies (e.g., setting a quit date, examining use patterns, reducing stress, and avoiding known triggers) or to self-help alone (manual only).[62] Prolonged abstinence (after a 30-day grace period) from all tobacco was significantly higher at 3 months (30.9% vs. 6.8%) and at 6 months (30.9% vs. 9.8%) in the quitline intervention compared with manual-only groups. The phone counseling again appears to be an important element in increasing quit rates. While self-help materials can be an important aid for supporting quitting, the effect of these programs is greatly enhanced with personal counseling. This model is being implemented by telephone helplines nationally. Mailed self-help materials can also support users to quit. While some helplines offer specific counseling and interventions to support ST users in quitting, there has been only one published study evaluating the cessation outcomes of telephone helpline programs with this population. Mushtaq and colleagues examined factors related to tobacco cessation among ST users registering for services with the Oklahoma Tobacco Helpline.[9] Of 959 male users who registered over an 8-year period, 374 completed the 7-month follow-up survey in which 43% of the participants reported 30-day abstinence from tobacco. These results suggest that state-sponsored quitlines may be a viable cessation resource for ST users interested in quitting.

Web-based

Web-based interventions have increased abstinence rates among ST users. In one RCT, 2523 US users were assigned to an enhanced or a basic website.[7] The enhanced ChewFree.com intervention included an interactive program, printable resources, and links to other websites, web forums, and education modules. The basic website consisted of static text based on a self-help quit guide.[31] Based upon the repeated point prevalence of all tobacco use at 3 and 6 months, the enhanced condition significantly increased tobacco abstinence rates compared to the basic condition (12.6% vs. 7.9%) using an intent-to-treat analysis. In a complete-case analysis, abstinence was 40.6% in the enhanced condition and 21.2% in the basic condition. Program exposure and use was significantly related to outcome as well as attrition. The conclusion was that tailored interactive web-assisted cessation programs can be effective for assisting adult ST users to quit.

Telephone versus web-based

There has been one study evaluating the efficacy of the State of California Quitline in conjunction with a web cessation program.[24] A 2 × 2 factorial design was used to examine the combined impact of access to a web-based intervention and quitline counseling for ST users who wanted to quit all tobacco use. They were randomly assigned to one of four study conditions: web only (n = 421), quitline only (n = 421), web + quitline (n = 417), or a self-help control (printed guide; n = 424). All participants were mailed the printed cessation guide that was routinely mailed to ST users who call the California Tobacco Chewers' Helpline. Participants assigned to the quitline condition were offered proactive calls by trained counselors from the California Tobacco Chewers' Helpline. Counselors followed an effective protocol used for smokers[63] that was adapted for ST users. They used a CATI (computer-assisted telephone interviewing) computer display that guided each call and the scheduling of follow-up calls. The initial call included discussion of motivation, quitting methods, previous quit attempts, social support, and setting a quit date. They used motivational interviewing[52] to boost readiness to quit and cognitive-behavioral techniques to increase self-efficacy and plan for challenging situations. They offered up to four follow-up calls to review the personalized plan, discuss relapse prevention, and encourage another quit attempt if needed. Other topics included health risks of ST use, nicotine withdrawal, and dual-tobacco use. Participants in the web + quitline condition were offered both the web content and quitline counseling. Counselors could access an online dashboard on their CATI display that enabled them to review participant use of the web intervention. They encouraged participants to use the web program during each call.

Participants in the web-only condition were given access to the fully automated, tailored, interactive enhanced web-based ST cessation intervention used in the ChewFree trial.[7] Program content emphasized cognitive-behavioral therapy (CBT) themes and related strategies—delivered as text, interactive activities, and videos. The program embodied a hybrid structure that combined information architecture design with some open access and some drill down activities[64] that emphasized three sequential phases (planning to quit, quitting, and staying quit).

This large RCT examined the combined impact of offering a web program and quitline counseling. Both interventions when offered alone yielded greater abstinence than the control (self-help booklet condition) and achieved levels of abstinence consistent with other studies. However, the authors found no additive or synergistic benefit of offering both the web and quitline, as has been observed for smoking cessation.[65] Although the abstinence rates for quitline counseling are substantially higher than those typically reported for smoking cessation,[66] the overall abstinence rates achieved in this trial for all groups—including the control—are consistent with the results obtained in earlier ST cessation studies that used low-intensity interventions.[7,23,33]

Counseling via telephone helpline

A recent study examined predictors related to tobacco abstinence among ST users seeking cessation assistance from the telephone helpline in Oklahoma.[9] Data were collected between 2004 and 2012 for 959 exclusive ST users, with follow-up at 7 months on 374 participants. At the follow-up, 43% reported a 30-day abstinence from ST, and each additional call increased the likelihood of cessation. Higher levels of motivation at baseline were associated with higher quit rates, but the use of nicotine replacement was not associated with abstinence.[9]

Oral substitutes for ST cessation

Herbal chew is a nicotine-free, nontobacco product available in US convenience stores or on the Internet. It comes as a chopped mint or other plant blend to be put in the mouth to replace the oral sensation of ST. Several studies have noted that oral substitutes can be helpful, as they provide a non-nicotine replacement that may aid in reducing craving symptoms. ST cessation guides suggest a wide range of products, including chewing gum, nuts, sunflower seeds, beef jerky, and cinnamon sticks.[31]

One study evaluated the efficacy of an herbal chew product (mint snuff) in a 2 × 2 design with 402 subjects randomized to the nicotine patch or placebo crossed with mint snuff or no mint snuff.[67] Mint snuff did not increase abstinence rates but significantly reduced craving and symptoms of withdrawal.

Although our review does not include studies of adolescents, it is notable that in one study with young ST users the authors reported that oral substitutes may be an important adjunctive aid for quitting.[68] Chakravorty assigned 70 rural male adolescent ST users (aged 14–18 years and averaging 1.5 dips per day) to one of three conditions: use of a nontobacco product composed of crushed mint leaves (mint snuff), use of nicotine gum, or attending a lecture-only control condition. Subjects in the mint snuff group were significantly more likely to report decreased use of ST than subjects in the other two conditions.[68] Oral substitutes can be an important element of assisting chewers in quitting their habit, and a variety of substitutes exist for this purpose.

Pharmacotherapy for ST cessation

Pharmacotherapies have been evaluated and approved by the FDA for smokers, but to date there has been no approval for any adjunctive aid for ST cessation. However, there is a developing research base on pharmacotherapy. These products include nicotine replacement therapy (gum, lozenge, and patch), bupropion sustained-release (SR), and varenicline (Table 5.4).

Nicotine replacement therapy

Nicotine replacement therapy is available by prescription and over the counter in both brand name and generic versions. It comes in a variety of delivery modes and doses. Studies indicate that it does not increase long-term (>6 months) abstinence rates among ST users. However, it does appear to decrease withdrawal and craving symptoms and increase short-term (10–12 weeks) abstinence rates. Treating withdrawal is important because ST users experience the same constellation of symptoms (craving, irritability, frustration, anger, difficulty concentrating, restlessness, impatience, increased appetite, and depressed mood) as smokers.

Nicotine gum

In a study evaluating the efficacy of the 2 mg gum for treatment of ST use, 210 adult users were randomized to 8 weeks of gum or placebo along with either the group behavioral intervention or minimal contact.[69] Gum did not significantly increase tobacco abstinence rates. However, it significantly decreased craving and nicotine withdrawal compared to placebo. Our clinical experience is that 2 mg gum does not provide enough nicotine replacement during withdrawal from ST, and therefore we highly recommend the

Table 5.4 Pharmacologic aids for smokeless tobacco cessation.

Study	Intervention	Design	Outcomes
Nicotine replacement therapy: Nicotine gum			
Hatsukami 1996[69]	2-mg nicotine gum for 8 weeks starting at 6 pieces per day then tapering with an option for a third month	RCT with 2 × 2 design; intensive counseling versus minimal contact crossed with nicotine gum versus placebo; 210 subjects	Gum significantly decreased tobacco withdrawal symptoms but did not increase abstinence from tobacco.
Nicotine replacement therapy: Nicotine lozenge			
Ebbert 2009[70]	4-mg nicotine lozenge for 12 weeks including tapering period	RCT; placebo-controlled; 270 subjects	Lozenge significantly increased abstinence from tobacco at 3 months and significantly decreased tobacco craving and withdrawal.
Ebbert 2010[71]	4 mg nicotine lozenge for 12 weeks mailed to subjects	RCT; placebo-controlled; 60 subjects	Nicotine lozenge did not increase abstinence from tobacco but significantly decreased tobacco withdrawal symptoms.
Nicotine replacement therapy: Nicotine patch			
Howard-Pitney 1999[72]	15-mg nicotine patch for 6 weeks	RCT; placebo-controlled; 410 subjects	Patch significantly increased ST abstinence at 3 months.
Hatsukami 2000[67]	21-mg nicotine patch for 10 weeks including tapering period	RCT with a 2 × 2 design; active versus placebo patch crossed with mint snuff versus none; 402 subjects	Patch significantly increased abstinence from tobacco at 10 and 15 weeks and significantly decreased tobacco craving and tobacco withdrawal symptoms.
Stotts 2003[73]	Nicotine patch tailored to baseline cotinine: > 150 ng/	RCT; placebo-controlled; 300 subjects	Patch did not increase abstinence from tobacco.

(Continued)

Table 5.4 Pharmacologic aids for smokeless tobacco cessation.—cont'd

Study	Intervention	Design	Outcomes
	mL received 21 mg initially, otherwise 14 mg with medication tapering for 6 weeks of treatment		
Ebbert 2007[74]	Nicotine patch: **1.** 63 mg **2.** 42 mg **3.** 21 mg **4.** Placebo	RCT; multidose, placebo-controlled; 42 subjects	Statistically significant dose—response relationship with higher nicotine doses was associated with less tobacco withdrawal.
		Bupropion	
Dale 2007[75]	Bupropion SR 150-mg by mouth twice a day for 12 weeks	RCT; placebo-controlled; 225 subjects	Bupropion SR did not increase abstinence from ST but significantly attenuated weight gain and significantly decreased tobacco craving.
Dale 2002[76]	Bupropion SR 150-mg by mouth twice a day for 12 weeks	RCT; placebo-controlled; 68 subjects	Bupropion nonsignificantly increased short-term tobacco abstinence rates and significantly decreased tobacco withdrawal.
		Varenicline	
Fagerström 2010[77]	Varenicline for 1 2 weeks	RCT; placebo-controlled; 431 subjects	Varenicline significantly increased abstinence from ST at 6 months.
Ebbert 2011[78]	Varenicline for 12 weeks	RCT; placebo-controlled; 76 subjects	Varenicline significantly decreased tobacco craving.

RCT = randomized controlled trial.

4 mg dose. There has not been a published study evaluating the efficacy of 4 mg gum, but in our interventions snuff users found more relief from withdrawal symptoms with 4 versus 2 mg.

Nicotine lozenge

In a study evaluating the efficacy of the 4 mg lozenge, 270 subjects were randomized to a 12-week tapering regimen of lozenges.[70] Compared to placebo, 4 mg significantly increased self-reported all-tobacco abstinence (44.1% vs. 29.1%) and self-reported ST abstinence (50.7% vs. 34.3%) compared with placebo at 12 weeks. The lozenge significantly decreased tobacco craving and nicotine withdrawal compared to placebo. In a small randomized pilot study (n = 60) evaluating the efficacy of mailing the 4 mg lozenge to ST users combined with phone support, the lozenge significantly decreased withdrawal symptoms compared to placebo.[71]

Nicotine patch

In a study evaluating the efficacy of the 15 mg per 16-hour patch, 410 adult ST users were randomized to the patch or placebo plus a behavioral intervention for 6 weeks.[72] All received two sessions with a pharmacist at baseline and at 4 weeks, as well as self-help materials and phone support at 48 hours and 10 days after the target quit date. The patch significantly increased abstinence rates at 3 months (31% vs. 25%), and less craving was observed at 48 hours after the target quit date. This program demonstrated the potential of using pharmacists as interventionists; other professional groups could expand the reach of cessation programs.

Hatsukami and colleagues evaluated the 21 mg per day patch for 6 weeks with a 4-week taper, compared to placebo; 402 subjects were randomized to patch or placebo plus mint snuff or no mint snuff.[67] Compared to placebo, the nicotine patch significantly increased tobacco abstinence rates at 10 weeks (67% vs. 53%) and at 15 weeks (52% vs. 43%). It also significantly decreased craving and withdrawal symptoms.

In another study, researchers evaluated high-dose nicotine patch therapy; 42 ST users were randomized to a 63 mg per day patch, a 42 mg per day patch, a 21 mg per day patch, or placebo.[74] Patches were given for 8 weeks, and all subjects received behavioral counseling. No significant differences were observed in abstinence rates between the four groups at 6 months. However, a significant relationship between higher patch doses and a greater degree of withdrawal symptom relief was observed, indicating that for very high levels of use (7 or more tins per week) it would be appropriate to use higher dose patches (two 21 mg patches per day).

Stotts and colleagues assessed the efficacy of nicotine patches in combination with behavioral therapy for treatment of ST addiction.[73] The study randomized 303 participants aged 14 to 19 to one of three conditions: control and two types of intervention. Participants in the control condition received a brief counseling session and follow-up call. Those in the two intervention conditions received a 6-week behavioral intervention, with one group getting active nicotine patches and the other getting placebo patches. The combined active treatment groups had higher cessation rates that the control condition; however, the active patch offered no improvement over placebo.

In sum, there is good support for the use of nicotine replacement products to reduce craving during the quitting process. Although nicotine lozenges appear to increase abstinence in controlled settings, in real-life settings most people do not use them often enough (i.e., according to directions) to get adequate replacement of the nicotine that they normally receive from their smokeless products. This suggests a need for further research on increasing medication compliance with this product. In addition, clinicians must work with ST users to determine the best type and dose of nicotine replacement therapy for their individual needs and provide instruction on proper use.

Bupropion

Bupropion is available in generic form only by prescription from a physician. It has not been demonstrated to increase short- or long-term abstinence rates among ST users, but it does decrease tobacco craving and reduce postcessation weight gain. In a study evaluating the efficacy of bupropion SR, 270 ST users were randomized to medication or placebo for 12 weeks.[75] Bupropion SR led to significantly less tobacco craving up to 14 days after the target quit date and less weight gain (1.7 ± 2.9 kg increase for bupropion, 3.2 ± 2.7 kg for placebo). Weight gain attenuation has also been observed in a smaller pilot study of bupropion SR for ST users[76]; the mean weight change from baseline to end of treatment was 0.7 ± 1.9 kg for bupropion and 4.4 ± 2.4 kg for placebo ($P = 0.03$).

Varenicline

Varenicline (trade names Chantix, Champix) is also available only by prescription. In a randomized placebo controlled study in 431 Scandinavian snus users, 1 mg by mouth twice daily for 12 weeks significantly increased continuous tobacco abstinence rates at weeks 9–12 (39% vs. 59%; $P < .001$) and at weeks 9–26 (34% vs. 45%; $P = 0.012$).[77] In a 12-week randomized placebo-controlled pilot study of 76 US ST users, varenicline significantly decreased tobacco craving.[78] However, this study was underpowered to assess abstinence outcomes.

ST reduction methods

Reduction-based cessation methods have shown mixed success among users who are motivated to quit. Jerome and colleagues conducted two small studies on computerized scheduled gradual reduction,[79] an approach modeled after Cinciripini and colleagues' scheduled smoking work,[80] which involves cueing smokers to smoke at progressively longer intervals until cessation is achieved. Jerome and colleagues adapted this approach using LifeSign, a credit card–sized computer designed for ST users, and found that it produced quit rates as high as 56% at the end of treatment, 29% at 3 months, and as high as 19% at 12-month follow-up.[79] No comparison arm was used in these studies. Severson and colleagues compared a gradual reduction approach using LifeSign to a self-help manual and video intervention. Self-reported sustained abstinence rates (quit at 2- and 6-month follow-up) were higher for those in the self-help manual group (24.5%) than in the gradual reduction group (18.4%). Participants in both groups received telephone support for their quit effort.[23]

Reduction-based methods also have been targeted at ST users who are not willing to quit. Published studies have examined (1) switching to lower nicotine brands,[81] (2) substitution of tobacco-free snuff,[67] and (3) medication-assisted reduction.[82,83] Brand switching and substitution of tobacco-free snuff lead to significant reductions in mean cotinine and NNAL [(4-(methylnitrosamino)-1-(3-pyridyl)-1-butanol)] concentrations.[81,84] Further, significant ST reductions from baseline to 12-week follow-up were also observed in the active intervention arms of the brand switching and substitution of tobacco-free snuff studies.[67] Medication-assisted reduction, including the addition of a nicotine lozenge to a behavioral ST reduction intervention, did not improve quit rates over behavioral intervention alone,[83] although both groups reported significant reductions in ST use from baseline to 12-week follow-up.[83] Pilot results for 12 weeks of varenicline with ST reduction resulted in 50% of participants reducing their use by 50% and 15% reporting abstinence at 12 weeks.[82] At 6-

month follow-up, reduction was still at 50% and abstinence was 10%. No comparison group was reported in this pilot study, limiting the interpretation of these results.[82]

One study compared immediate ("cold turkey") quitting to gradual reduction among ST users not willing to quit.[85] The immediate quit approach was more effective than gradual reduction in achieving cessation at 12- and 26-week follow-up.[85] Nonquitters in both groups reported significant reductions in ST use, with no difference between groups. Another study compared the nicotine lozenge with ST reduction to tobacco-free snuff with ST reduction.[86] There were no significant differences between groups. However, both groups significantly reduced ST use from baseline to 26-week follow-up, with similar biochemically confirmed abstinence rates at 26 weeks, suggesting that both combinations of methods are effective.[86]

Recommendations

Research has been done on a wide range of interventions to promote ST cessation. Programs for adult users have shown encouraging evidence for dental office interventions and clinical interventions involving multiple sessions and counselor support. Feedback on oral exams and phone counseling appear to be key elements for success. One limitation of dental programs is that many high-risk users do not see a dentist.[87] Nonetheless, we recommend that all oral health professionals be engaged in the prevention and treatment of ST dependence. The American Dental Association and the American Dental Hygiene Association recommend that their members routinely discuss tobacco use with their patients.[88] However, oral health professionals need to be trained to recognize oral disease caused by ST, and to provide brief tobacco use interventions in the context of oral healthcare. If they are not comfortable with the treatment options, they should have systems in place to refer patients who want treatment to specialists with the necessary training. Systems that allow for routine delivery of models such as "Ask-Advise-Refer" or "Ask-Advise-Connect"[89] should be adopted and implemented in all types of dental practices. Reimbursement models similar to those in medical settings should also be implemented for tobacco cessation counseling.

Nicotine replacement (patches, gum, or lozenges) can help reduce withdrawal symptoms and craving, but they are ineffective for increasing long-term tobacco abstinence. Bupropion and varenicline decrease craving among ST users trying to quit, but require a physician's prescription. Varenicline increases short- and long-term tobacco abstinence rates. We recommend that, where available, medication could be used to reduce symptoms associated with quitting, and to increase short-term and long-term quit rates in the case of varenicline.

Interventions for special populations of users (e.g., Native Americans, baseball players) have been developed and evaluated and should be used when possible. Cultural adaptations are needed to provide interventions that are appropriate for both the context of use and the product being used. Consider tailoring materials to special populations of chewers and dippers. For example, there are special versions of the Enough Snuff self-help guide for Native American, military, and Spanish-speaking populations of chewers and dippers. Colleagues are in the process of developing versions of self-help guides for specific occupations that have high levels of use, such as firefighters.

In resource-constrained environments or where cessation counselors or clinics are not available, telephone coaching is a cost-effective strategy. Self-help manuals, especially when used in conjunction with telephone coaching, also can increase access to effective programs. Web-based programs also appear to be low-cost and effective. There is good support for their effectiveness, and they greatly increase the reach of intervention. Since ST use is more prevalent in rural areas where medical, dental, and mental health resources may be scarce, the use of the Internet as a delivery system for self-help programs is both warranted and supported by research. The increased access to the Internet throughout the United States supports the use of web-based interventions.[90]

Future directions

While there has been some support for gradual reduction methods to help ST users quit, more research is needed. Recent efforts to apply scheduled gradual reduction to help chewers have received support in pilot studies, though full-scale evaluations have not been completed. An innovative approach to test scheduled gradual reduction was evaluated in a pilot study by Noonan and colleagues at Duke University. Based on their successful experience using text messages for smoking cessation[91], they randomized participant chewers from rural and medically underserved areas to receive either scheduled gradual reduction plus text messages or text messages alone. Each participant received the intervention via their cell phone, and preliminary data support this approach.[92] The study also assessed the impact of text messaging alone and found support for this approach as well.[92]

Mobile phone apps for ST cessation have not been adequately evaluated, but this is an area where research could demonstrate another avenue for the delivery of self-help interventions. Although text messaging has been shown to be efficacious for smokers,[93–95] it has not yet been evaluated for ST users. It seems quite logical that this approach would be viable. There is a program of text messaging for ST cessation supported by the National Cancer Institute, available as a free downloadable app from the smokefree.gov website,[96] but it has not been evaluated. The Veterans Administration supports a resource entitled SmokefreeVET, a text message program for ST users[96]; however, there has been no evaluation to date. Given the evidence to support web-based interventions,[7,24] it is likely that similar programs offered via a mobile app would be effective. Web-based programs could be adapted as apps to increase usability in rural areas or other locations where Internet access is not reliable or readily available to increase reach to those ST users.

One theory-driven approach that has shown promise for development is cognitive dissonance induction. It is based on the premise that individuals strive toward consistency between their beliefs and behaviors.[97] When inconsistencies between attitudes, beliefs, or behaviors occur, individuals tend to adjust their beliefs or behaviors to reduce the discomfort of the dissonance. Reduction of discomfort can be achieved in one of three ways: (1) adjusting one or more attitudes or behaviors to make the relationship between the two consistent, (2) acquiring new information that offsets the dissonant belief, or (3) reducing the importance of the attitudes or beliefs so that the inconsistent behavior can continue.

Interventions that use this method to elicit behavior change do so through situations that create high levels of dissonance. Regardless of the behavior being targeted, there are several common concepts that are central to effectively eliciting cognitive dissonance induction. First, participation in the induction activity must be voluntary. Voluntary participation allows for the individual to attribute the inconsistency between beliefs and behaviors as existing within themselves, rather than being due to the demands of a given situation.[97] Effortful involvement (i.e., actively engaging in treatment exercises) is also required and is believed to result in greater dissonance and greater motivation for change.[98] And finally, public statements of beliefs (i.e., counterattitudinal advocacy) are thought to elicit heightened dissonance responses.[99]

Of particular importance, cognitive dissonance approaches have also shown success in treating substance use,[100,101] reducing the initiation of smoking,[102] and improving short-term outcomes for smoking cessation.[103–105] Simmons and colleagues[103] demonstrated greater quit rates at 1-month follow-up for smokers who engaged in videotaped discussions of the consequences of smoking delivered via computer. Although these results are promising, no long-term significant outcomes were found, which suggest that increased exposure to dissonance induction activities (dosage) and reach of cognitive dissonance induction approaches might yield improved abstinence. While this approach has not been evaluated with ST users, it offers a unique avenue to further study.

A different approach to increasing ST cessation focuses on enlisting the users' wives and domestic partners. During recruitment for a self-help study[61] we observed many women attempting to enroll their husband or partner. We decided to explore the possibility of intervening directly with women as a means of reaching users who may not themselves be thinking of quitting.

Social support has an important role in tobacco cessation. Supporters can motivate cessation by expressing concerns about health effects and influences on children; they can also help buffer the stress of withdrawal and facilitate access to cessation resources.[106] Previous research with supporters of smokers had focused on increasing positive behaviors (e.g., helping the smoker learn about quitting programs) and decreasing negative behaviors (e.g., nagging the smoker to quit or monitoring their potential use during a quit attempt). In our first study with supporters, we recruited 522 women, providing them with a printed booklet and giving them access to the ChewFree.com web-based program, which could be shared with the ST user. At 6 weeks, the users' cessation rates were 10.9% for intervention versus 5.1% for a control group that simply was given access to ChewFree.com with generic support tips.

In our next study with supporters, we switched to a new theoretical framework. "Perceived partner responsiveness" is based on the empirical finding that support is better received if the supporter conveys respect, understanding, and caring to the recipient. This framework can potentially help supporters understand which behaviors are positive and which are negative through the three stages of the cessation process: deciding and preparing to quit, quitting (and going through the associated withdrawal from nicotine), and maintenance. The intervention consists of a website and mailed booklet. The website suggests that women encourage their partner to use ChewFree.com or call a tobacco quitline if he decides to make a quit attempt. We recruited 1145 women and randomized them to the intervention or a delayed treatment control, asking them to provide a surrogate report of their partners' cessation activities at 6-week and 7.5-month follow-up. For partners of women completing the intervention, 12.4% had quit all tobacco at 7.5 months, compared with 6.6% for delayed treatment. Further, change in responsiveness-based and instrumental behaviors at 6 weeks independently mediated cessation at 7.5 months, and change in responsiveness-based attitudes indirectly mediated cessation outcomes through their effect on responsiveness-based behaviors, confirming the value of the responsiveness-based approach.[107]

Conclusions

There has been extensive research evaluating a wide range of interventions to assist ST users to quit. It has yielded good evidence and support for several methods, with some outcomes exceeding those for smoking cessation. However, we still have much to learn about the best ways to promote ST cessation through institutions or with individuals. Despite public health campaigns to reduce the use of snuff and chew, a significant number of Americans continue to be regular users. Tobacco companies have consolidated their holdings to include ST and innovative products, and a vast array of flavors appears to maintain a large group of regular users. We have identified several promising avenues for intervention and further research, especially tailored programs for special populations among whom ST use is particularly high. We are encouraged to see the large body of research, but much more could be done to promote both the use of existing effective interventions and the development of innovative programs.

References

1. Lipari RN, Van Horn SL. Trends in smokeless tobacco use and initiation: 2002 to 2014. In: *The CBHSQ Report: May 31, 2017*. Rockville, MD: Center for Behavioral Health Statistics and Quality, Substance Abuse and Mental Health Services Administration; 2017.
2. U.S. Department of Health and Human Services. In: *The Health Consequences of Smoking: 50 Years of Progress: A Report of the Surgeon General*. Atlanta, GA: U.S. Department of Health and Human Services, Centers for Disease Control and Prevention, National Center for Chronic Disease Prevention and Health Promotion, Office on Smoking and Health; 2014.
3. U.S. Department of Health and Human Services. In: *Preventing Tobacco Use Among Youth and Young Adults: A Report of the Surgeon General*. Atlanta, GA: Centers for Disease Control and Prevention, National Center for Chronic Disease Prevention and Health Promotion, Office on Smoking and Health; 2012.
4. National Cancer Institute. In: *Smokeless Tobacco and Public Health: A Global Perspective*. Bethesda, MD: US department of health and human services, centers for disease control and prevention and national institutes of health, national cancer institute; 2014. NIH publication no. 14—7983.

5. Ebbert JO, Elrashidi MY, Stead LF. Interventions for smokeless tobacco use cessation. *Cochrane Database Syst Rev*. 2015;(10). Cd004306.
6. Biener L, Abrams DB. The contemplation ladder: validation of a measure of readiness to consider smoking cessation. *Health Psychol*. 1991;10(5):360–365.
7. Severson HH, Gordon JS, Danaher BG, Akers L. ChewFree.com: evaluation of a web-based cessation program for smokeless tobacco users. *Nicotine Tob Res*. 2008; 10(2):381–391.
8. Severson HH, Peterson AL, Andrews JA, et al. Smokeless tobacco cessation in military personnel: a randomized controlled trial. *Nicotine Tob Res*. 2009;11(6):730–738.
9. Mushtaq N, Boeckman LM, Beebe LA. Predictors of smokeless tobacco cessation among telephone quitline participants. *Am J Prev Med*. 2015;48(1):S54–S60.
10. Fagerstrom KO, Schneider NG. Measuring nicotine dependence: a review of the fagerstrom tolerance questionnaire. *J Behav Med*. 1989;12(2):159–182.
11. Boyle RG, Jensen J, Hatsukami DK, Severson HH. Measuring dependence in smokeless tobacco users. *Addict Behav*. 1995;20(4):443–450.
12. Roins LN, Cottler LB, Bucholz KK, Compton WM, North CS, Rourke KM. In: *NIMH Diagnostic Interview Schedule, Version IV*. St. Louis, MO: Washington University; 1999.
13. Ebbert JO, Severson HH, Danaher BG, Schroeder DR, Glover ED. A comparison of three smokeless tobacco dependence measures. *Addict Behav*. 2012;37(11): 1271–1277.
14. Glover ED, Nilsson F, Westin A, Glover PN, Laflin MT, Persson B. Developmental history of the Glover-Nilsson smoking behavioral questionnaire. *Am J Health Behav*. 2005;29(5):443–455.
15. Ebbert JO, Patten CA, Schroeder DR. The Fagerström test for nicotine dependence-smokeless tobacco (FTND-ST). *Addict Behav*. 2006;31(9):1716–1721.
16. Piper ME, Piasecki TM, Federman EB, et al. A multiple motives approach to tobacco dependence: the Wisconsin Inventory of Smoking Dependence Motives (WISDM-68). *J Consult Clin Psychol*. 2004;72(2):139.
17. Mushtaq N, Beebe LA, Vesely SK, Neas BR. A multiple motive/multi-dimensional approach to measure smokeless tobacco dependence. *Addict Behav*. 2014;39(3): 622–629.
18. Mushtaq N, Beebe LA. Evaluating the role of smokeless tobacco use indices as brief measures of dependence. *Addict Behav*. 2017;69:87–92.
19. Mushtaq N, Beebe LA. Assessment of the tobacco dependence screener among smokeless tobacco users. *Nicotine Tob Res*. 2016;18(5):885–891.
20. Severson HH, Gordon JS. *Enough Snuff: A Guide for Quitting Smokeless Tobacco*. Eugene, OR: Applied Behavior Science Press; 2000.
21. Severson HH. Smokeless tobacco: risk, epidemiology and cessation. In: *Nicotine Addiction: Principles and Management*. New York: NY: Oxford University Press; 1993: 262–278.
22. Hughes JR, Keely JP, Niaura RS, Ossip-Klein DJ, Richmond RL, Swan GE. Measures of abstinence in clinical trials: issues and recommendations. *Nicotine Tob Res*. 2003;5(1):13–25.
23. Severson HH, Akers L, Andrews JA, Lichtenstein E, Jerome A. Evaluating two self-help interventions for smokeless tobacco cessation. *Addict Behav*. 2000;25(3):465–470.
24. Danaher BG, Severson HH, Zhu S-H, et al. Randomized controlled trial of the combined effects of web and quitline interventions for smokeless tobacco cessation. *Internet Interv*. 2015;2(2):143–151.
25. Severson HH. What have we learned from 20 years of research on smokeless tobacco cessation? *Am J Med Sci*. 2003;326(4):206–211.

26. Kalkhoran S, Grana RA, Neilands TB, Ling PM. Dual use of smokeless tobacco or e-cigarettes with cigarettes and cessation. *Am J Health Behav*. 2015;39(2):277–284.
27. Glover ED, Schroeder KL, Henningfield JE, Severson HH, Christen AG. An interpretative review of smokeless tobacco research in the United States: Part I. *J Drug Educ*. 1988;18(4):285–310.
28. Severson HH, Eakin EG, Stevens VJ, Lichtenstein E. Dental office practices for tobacco users: independent practice and HMO clinics. *Am J Public Health*. 1990; 80(12):1503–1505.
29. Hoffman D, Brunnemann KD, Adams JD, Hecht SS. Laboratory studies on snuff-dipping and oral cancer. *Cancer J*. 1986;1(1):9.
30. Fagerström K, Eissenberg T. Dependence on tobacco and nicotine products: a case for product-specific assessment. *Nicotine Tob Res*. 2012;14(11):1382–1390.
31. Severson HH, Gordon JS. *Enough Snuff: A Guide for Quitting Smokeless Tobacco*. 8th ed. Scotts Valley CA: ETR Associates; 2010.
32. West R, McNeill A, Raw M. Smokeless tobacco cessation guidelines for health professionals in England. *Br Dent J*. 2004;196(10):611–618.
33. Ebbert J, Montori VM, Erwin PJ, Stead LF. Interventions for smokeless tobacco use cessation. *Cochrane Database Syst Rev*. 2011;(2).
34. Bergman HE, Hunt YM, Augustson E. Smokeless tobacco use in the United States military: a systematic review. *Nicotine Tob Res*. 2012;14(5):507–515.
35. Gordon JS, Severson HH. Tobacco cessation through dental office settings. *J Dent Educ*. 2001;65(4):354–363.
36. Stevens VJ, Severson H, Lichtenstein E, Little SJ, Leben J. Making the most of a teachable moment: a smokeless-tobacco cessation intervention in the dental office. *Am J Public Health*. 1995;85(2):231–235.
37. Severson HH, Walker HM, Nave G. Systematic screening for at risk students: a multiple gating approach. In: *The Oregon Conference Monograph*. 1998:80–85.
38. Andrews JA, Severson HH, Lichtenstein E, Gordon JS, Barckley MF. Evaluation of a dental office tobacco cessation program: effects on smokeless tobacco use. *Ann Behav Med*. 1999;21(1):48–53.
39. U.S. Department of Health and Human Services. In: *Treating Tobacco Use and Dependence: 2008 Update*. Rockville, MD: Agency for Healthcare Research and Quality; 2008.
40. Orleans CT. Increasing the demand for and use of effective smoking-cessation treatments: reaping the full health benefits of tobacco-control science and policy gains—in our lifetime. *Am J Prev Med*. 2007;33(6):S340–S348.
41. Gordon JS, Andrews JA, Lichtenstein E, Severson HH, Akers L. Disseminating a smokeless tobacco cessation intervention model to dental hygienists: a randomized comparison of personalized instruction and self-study methods. *Health Psychol*. 2005; 24(5):447.
42. Severson HH, Andrews JA, Lichtenstein E, Gordon JS, Barckley MF. Using the hygiene visit to deliver a tobacco cessation program: results of a randomized clinical trial. *J Am Dent Assoc*. 1998;129(7):993–999.
43. Gordon JS, Andrews JA, Crews KM, Payne TJ, Severson HH, Lichtenstein E. Do faxed quitline referrals add value to dental office-based tobacco-use cessation interventions? *J Am Dent Assoc*. 2010;141(8):1000–1007.
44. Danesh D, Paskett ED, Ferketich AK. Disparities in receipt of advice to quit smoking from health care providers: 2010 National Health Interview Survey. *Prev Chronic Dis*. 2014;11. E131-E131.
45. Agaku IT, Ayo-Yusuf OA, Vardavas CI. A comparison of cessation counseling received by current smokers at US dentist and physician offices during 2010–2011. *Am J Public Health*. 2014;104(8):e67–e75.

46. Shelley D, Wright S, McNeely J, et al. Reimbursing dentists for smoking cessation treatment: views from dental insurers. *Nicotine Tob Res.* 2012;14(10):1180–1186.
47. McMenamin SB, Halpin HA, Shade SB. Trends in employer-sponsored health insurance coverage for tobacco-dependence treatments. *Am J Prev Med.* 2008;35(4): 321–326.
48. Olmsted KL, Bray RM, Reyes-Guzman CM, Williams J, Kruger H. Overlap in use of different types of tobacco among active duty military personnel. *Nicotine Tob Res.* 2011;13(8):691–698.
49. McClellan SF, Olde BA, Freeman DH, Mann WF, Rotruck JR. Smokeless tobacco use among military flight personnel: a survey of 543 aviators. *Aviat Space Environ Med.* 2010;81(6):575–580.
50. Klesges RC, DeBon M, Vander Weg MW, et al. Efficacy of a tailored tobacco control program on long-term use in a population of U.S. military troops. *J Consult Clin Psychol.* 2006;74(2):295–306.
51. Cigrang JA, Severson HH, Peterson AL. Pilot evaluation of a population-based health intervention for reducing use of smokeless tobacco. *Nicotine Tob Res.* 2002;4(1):127–131.
52. Miller WR, Rollnick S. *Motivational Interviewing: Helping People Change.* Guilford press; 2012.
53. Miller WR, Rollnick S. *Motivational Interviewing: Preparing People for Change.* 2nd ed. New York, NY, US: Guilford Press; 2002.
54. Severson HH, Gordon JS. *Enough Snuff: Quitting Smokeless Tobacco Guide for Military Personnel.* Eugene, OR: Applied Behavior Science Press; 2003.
55. Severson HH, Gordon JS, Christiansen S. *Tough Enough to Quit.* Applied Behavior Science Press; 2003.
56. Ernster VL, Grady DG, Greene JC, et al. Smokeless tobacco use and health effects among baseball players. *J Am Med Assoc.* 1990;264(2):218–224.
57. Severson HH, Klein K, Lichtensein E, Kaufman N, Orleans CT. Smokeless tobacco use among professional baseball players: survey results, 1998 to 2003. *Tob Control.* 2005;14(1):31.
58. Walsh MM, Hilton JF, Masouredis CM, Gee L, Chesney MA, Ernster VL. Smokeless tobacco cessation intervention for college athletes: results after 1 year. *Am J Public Health.* 1999;89(2):228–234.
59. Gansky SA, Ellison JA, Rudy D, et al. Cluster-randomized controlled trial of an athletic trainer-directed spit (smokeless) tobacco intervention for collegiate baseball athletes: results after 1 year. *J Athl Train.* 2005;40(2):76–87.
60. Greene JC, Walsh MM, Masouredis C. A program to help major league baseball players quit using spit tobacco. *J Am Dent Assoc.* 1994;125(5):559–568.
61. Severson HH, Andrews JA, Lichtenstein E, Gordon JS, Barckley M, Akers L. A self-help cessation program for smokeless tobacco users: comparison of two interventions. *Nicotine Tob Res.* 2000;2(4):363–370.
62. Boyle RG, Enstad C, Asche SE, et al. A randomized controlled trial of telephone counseling with smokeless tobacco users: the ChewFree Minnesota study. *Nicotine Tob Res.* 2008;10(9):1433–1440.
63. Zhu SH, Stretch V, Balabanis M, Rosbrook B, Sadler G, Pierce JP. Telephone counseling for smoking cessation: effects of single-session and multiple-session interventions. *J Consult Clin Psychol.* 1996;64(1):202–211.
64. Danaher BG, McKay HG, Seely JR. The information architecture of behavior change websites. *J Med Internet Res.* 2005;7(2):1–10.
65. Graham AL, Cobb NK, Papandonatos GD, et al. A randomized trial of Internet and telephone treatment for smoking cessation. *Arch Intern Med.* 2011;171(1):46–53.
66. Stead LF, Hartmann-Boyce J, Perera R, Lancaster T. Telephone counselling for smoking cessation. *Cochrane Database Syst Rev.* 2013;(8). Cd002850.

67. Hatsukami DK, Grillo M, Boyle R, et al. Treatment of spit tobacco users with transdermal nicotine system and mint snuff. *J Consult Clin Psychol.* 2000;68(2):241.
68. Chakravorty BJ. A product substitution approach to adolescent smokeless tobacco cessation. *Diss Abstr.* 1992;53:2808–2809.
69. Hatsukami D, Jensen J, Allen S, Grillo M, Bliss R. Effects of behavioral and pharmacological treatment on smokeless tobacco users. *J Consult Clin Psychol.* 1996;64(1):153.
70. Ebbert JO, Severson HH, Croghan IT, Danaher BG, Schroeder DR. A randomized clinical trial of nicotine lozenge for smokeless tobacco use. *Nicotine Tob Res.* 2009; 11(12):1415–1423.
71. Ebbert JO, Severson HH, Croghan IT, Danaher BG, Schroeder DR. A pilot study of mailed nicotine lozenges with assisted self-help for the treatment of smokeless tobacco users. *Addict Behav.* 2010;35(5):522–525.
72. Howard-Pitney B, Killen JD, Fortmann SP. Quitting chew: results from a randomized trial using nicotine patches. *Exp Clin Psychopharmacol.* 1999;7(4):362–371.
73. Stotts R, Roberson P, Hanna E, Jones S, Smith C. A randomised clinical trial of nicotine patches for treatment of spit tobacco addiction among adolescents. *Tob Control.* 2003;12(suppl 4):iv11–iv15.
74. Ebbert JO, Dale LC, Patten CA, et al. Effect of high-dose nicotine patch therapy on tobacco withdrawal symptoms among smokeless tobacco users. *Nicotine Tob Res.* 2007; 9(1):43–52.
75. Dale LC, Ebbert JO, Glover ED, et al. Bupropion SR for the treatment of smokeless tobacco use. *Drug Alcohol Depend.* 2007;90(1):56–63.
76. Dale LC, Ebbert JO, Schroeder DR, et al. Bupropion for the treatment of nicotine dependence in spit tobacco users: a pilot study. *Nicotine Tob Res.* 2002;4(3):267–274.
77. Fagerström K, Gilljam H, Metcalfe M, Tonstad S, Messig M. Stopping smokeless tobacco with varenicline: randomised double blind placebo controlled trial. *BMJ.* 2010; 341:c6549.
78. Ebbert JO, Croghan IT, Severson HH, Schroeder DR, Hays JT. A pilot study of the efficacy of varenicline for the treatment of smokeless tobacco users in Midwestern United States. *Nicotine Tob Res.* 2011;13(9):820–826.
79. Jerome A, Fiero PL, Behar A. Computerized scheduled gradual reduction for smokeless tobacco cessation: development and preliminary evaluation of a self-help program. *Comput Hum Behav.* 2000;16(5):496–505.
80. Cinciripini PM, Wetter DW, McClure JB. Scheduled reduced smoking: effects on smoking abstinence and potential mechanisms of action. *Addict Behav.* 1997;22(6): 759–767.
81. Hatsukami DK, Ebbert JO, Anderson A, Lin H, Le C, Hecht SS. Smokeless tobacco brand switching: a means to reduce toxicant exposure? *Drug Alcohol Depend.* 2007; 87(2–3):217–224.
82. Ebbert JO, Croghan IT, North F, Schroeder DR. A pilot study to assess smokeless tobacco use reduction with varenicline. *Nicotine Tob Res.* 2010;12(10):1037–1040.
83. Ebbert JO, Edmonds A, Luo X, Jensen J, Hatsukami DK. Smokeless tobacco reduction with the nicotine lozenge and behavioral intervention. *Nicotine Tob Res.* 2010; 12(8):823–827.
84. Hatsukami DK, Ebbert JO, Edmonds A, et al. Smokeless tobacco reduction: preliminary study of tobacco-free snuff versus no snuff. *Nicotine Tob Res.* 2008;10(1):77–85.
85. Schiller KR, Luo X, Anderson AJ, Jensen JA, Allen SS, Hatsukami DK. Comparing an immediate cessation versus reduction approach to smokeless tobacco cessation. *Nicotine Tob Res.* 2012;14(8):902–909.
86. Ebbert JO, Severson HH, Croghan IT, Danaher BG, Schroeder DR. Comparative effectiveness of the nicotine lozenge and tobacco-free snuff for smokeless tobacco reduction. *Addict Behav.* 2013;38(5):2140–2145.

87. Andrews JA, Severson HH, Lichtenstein E, Gordon JS. Relationship between tobacco use and self-reported oral hygiene habits. *J Am Dent Assoc.* 1998;129(3):313–320.
88. American Dental Association. *Oral Health Topics: Smoking and Tobacco Cessation*; 2018. https://www.ada.org/en/member-center/oral-health-topics/smoking-and-tobacco-cessation.
89. Vidrine JI, Shete S, Cao Y, et al. Ask-Advise-Connect: a new approach to smoking treatment delivery in health care settings. *JAMA Intern Med.* 2013;173(6):458–464.
90. Hitlin P. *Internet, Social Media Use and Device Ownership in U.S. Have Plateaued after Years of Growth.* Pew Research Factank: News in the Numbers; 2018. http://www.pewresearch.org/fact-tank/2018/09/28/internet-social-media-use-and-device-ownership-in-u-s-have-plateaued-after-years-of-growth/.
91. Noonan D, Silva S, Njuru J, et al. Feasibility of a text-based smoking cessation intervention in rural older adults. *Health Educ Res.* 2018;33(1):81–88.
92. Identifier NCT02613689 ClinicalTrialsgov [Internet]. *Addressing Tobacco Use Disparities through an Innovative Mobile Phone Intervention: The Text to Forgo Smokeless Tobacco*; February 29, 2006. https://clinicaltrials.gov/ct2/show/NCT02613689, 2015 Nov 24.
93. Whittaker R, McRobbie H, Bullen C, Rodgers A, Gu Y. *Mobile Phone—based Interventions for Smoking Cessation.* The Cochrane Library; 2016.
94. Haskins BL, Lesperance D, Gibbons P, Boudreaux ED. A systematic review of smartphone applications for smoking cessation. *Transl Beh Med.* 2017;7(2):292–299.
95. Regmi K, Kassim N, Ahmad N, Tuah NAA. Effectiveness of mobile apps for smoking cessation: A review. *Tob Prev Cessation.* 2017;3(April).
96. *Smokefree.gov. Smokefreevet*; December 14, 2018. https://smokefree.gov/veterans.
97. Festinger L. *A Theory of Cognitive Dissonance.* Stanford, CA: Stanford University Press; 1957.
98. Green M, Scott N, Diyankova I, Gasser C. Eating disorder prevention: an experimental comparison of high level dissonance, low level dissonance, and no-treatment control. *Eat Disord.* 2005;13(2):157–169.
99. Aronson E, Fried C, Stone J. Overcoming denial and increasing the intention to use condoms through the induction of hypocrisy. *Am J Public Health.* 1991;81(12):1636–1638.
100. Barnett LA, Far JM, Mauss AL, Miller JA. Changing perceptions of peer norms as a drinking reduction program for college students. *J Alcohol Drug Educ.* 1996.
101. Ulrich J. A motivational approach to the treatment of alcoholism in the Federal Republic of Germany. *Alcohol Treat Q.* 1991;8(2):83–92.
102. Killen JD. Prevention of adolescent tobacco smoking: the social pressure resistance training approach. *JCPP (J Child Psychol Psychiatry).* 1985;26(1):7–15.
103. Simmons VN, Heckman BW, Fink AC, Small BJ, Brandon TH. Efficacy of an experiential, dissonance-based smoking intervention for college students delivered via the Internet. *J Consult Clin Psychol.* 2013;81(5):810.
104. Simmons VN, Webb MS, Brandon TH. College-student smoking: an initial test of an experiential dissonance-enhancing intervention. *Addict Behav.* 2004;29(6):1129–1136.
105. Simmons VN, Brandon TH. Secondary smoking prevention in a university setting: a randomized comparison of an experiential, theory-based intervention and a standard didactic intervention for increasing cessation motivation. *Health Psychol.* 2007;26(3):268.
106. Cohen S, Lichtenstein E, Mermelstein R, Kingsolver K, Baer J, Kamarck T. Social support interventions for smoking cessation. In: Gottlieb GH, ed. *Creating Support Groups: Formats, Processes and Effects.* New York: NY: Sage; 1988.
107. Akers L, Andrews JA, Lichtenstein E, Severson HH, Gordon JS. Effect of a responsiveness-based support intervention on smokeless tobacco cessation: the UCare-ChewFree randomized clinical trial. *Nicotine Tob Res.* 2019. https://doi.org/10.1093/ntr/ntz074.

CHAPTER SIX

Chemical characterization of smokeless tobacco products and relevant exposures in users

Irina Stepanov, Dorothy K. Hatsukami
University of Minnesota, Minneapolis, MN, United States

Introduction

Smokeless tobacco (ST) products are characterized by enormous diversity of types and formulations, from uncured dry tobacco that is used by itself to complex mixtures made with various additional ingredients that can modify tobacco addictiveness, toxicity, and carcinogenicity. Adding to this complexity is the diversity of tobacco plant types and the approaches to its processing that are used worldwide. Data on exposure to chemical toxicants and carcinogens from the use of ST are scarce and mostly limited to the United States. However, cancer risks associated with ST use worldwide mirror the variations in the levels of key known carcinogens in products. In India, many ST products contain high levels of certain carcinogens, and the incidence of head and neck cancers is remarkably high, while Swedish snus contains low levels of many harmful constituents and the associated cancer risk is virtually nonexistent. Therefore, chemical characterization of ST products and related exposures in users is key to the development of preventive measures such as consumer education and product regulation.

This chapter provides an overview of sources, levels, and variation of harmful constituents in ST products, relevant biomarkers to assess users' exposure to these constituents, and analytical methods to measure such constituents and their biomarkers in users. Variations in the levels of key harmful constituents across ST products and the impact of these variations on related exposures are emphasized, and policy implications of such variations are discussed at the conclusion of the chapter.

Smokeless Tobacco Products
ISBN: 978-0-12-818158-4
https://doi.org/10.1016/B978-0-12-818158-4.00006-6

Chemical diversity of ST products

Even though ST use does not involve burning, it exposes users to a highly complex chemical mixture. In addition to more than 3000 compounds that have been identified in unburned tobacco,[1] certain constituents can be formed during tobacco processing or introduced with other ingredients during product formulation.[2,3] Some of them are potent toxicants and carcinogens, and are responsible for addiction and chronic diseases caused by ST use.[2,4] For instance, nicotine, the major known addictive constituent in tobacco, is present in the green tobacco plant, and its levels depend on tobacco type.[5] Metals and inorganic salts such as nitrates and nitrites are also present in the plant; levels are influenced by soil composition.[4,6–8] Tobacco processing is the key step in the formation of tobacco-specific *N*-nitrosamines (TSNA), a major group of carcinogens in tobacco products.[2,9] These compounds are formed via nitrosation of tobacco-specific alkaloids, and the formed levels depend on tobacco type, levels of nitrates and nitrites in the tobacco, and the approaches to processing and storage.[6,10,11] Polycyclic aromatic hydrocarbons (PAH) and volatile aldehydes, other important harmful agents, are examples of constituents that are introduced as contaminants or by-products during tobacco processing or with nontobacco ingredients during product formulation.[4,12,13] Variations of these and other constituents across ST products are discussed later, and approaches to reduction of some key harmful toxicants and carcinogens are discussed in Chapter 8. Note that comparison of constituent levels in ST products and cigarettes (either tobacco filler or smoke) would not be meaningful because of the difference in the route of administration. Because of the large surface area in the lung, constituent absorption from cigarette smoke is much more efficient than from ST in the oral cavity. Biomarkers of tobacco constituent exposure account for the route of intake and other factors such as pattern of use and therefore are more informative in comparing the toxic and carcinogenic potential of ST and cigarettes. Biomarkers will be discussed in a separate section later in this chapter.

Nicotine and unprotonated nicotine

Nicotine content and its bioavailability play a key role in abuse liability of ST and users' continuous exposure to other harmful constituents. Being a weak base, nicotine can exist in three forms: unprotonated, monoprotonated, and diprotonated. Unprotonated (also referred to as "free") is the biologically available form that can easily cross cellular membranes and quickly reach the brain, and products with higher levels of unprotonated nicotine can be more addictive.[14,15] Relative proportion of total nicotine content in this biologically available form can be determined from the Henderson–Hasselbalch equation, based on the measured total nicotine, pH values, and a pKa of 8.02 for nicotine.[16] Natural tobacco pH is slightly acidic, leading to nearly 100% of nicotine being present in the monoprotonated form that cannot be absorbed efficiently. However, even slight variations in pH lead to substantial shifts in unprotonated content. This is illustrated in Fig. 6.1, which shows the shift between protonated and unprotonated nicotine within the range of pH values measured in ST products worldwide.

Because of the slightly acidic pH of tobacco, adding alkaline ingredients to ST is nearly ubiquitous, and this leads to wide variation in bioavailable nicotine levels across products with the same amount of total nicotine.[2,20] Modifications are known to be done by both the manufacturers and, in the case of custom-prepared products in South-East Asia (see Chapter 2 for the WHO definition of this region), by users. For instance, US manufacturers manipulated product pH in the past, tailoring the amount of bioavailable nicotine in moist snuff brands to novice or addicted users as part of a "graduation strategy."[21] There is still substantial variation in unprotonated levels in US moist snuff, as can be seen from the analysis of the top 40 brands by Richter et al.[20] Fig. 6.2 illustrates unprotonated nicotine levels in 38 brands from that paper, which had total nicotine in relatively narrow ranges; with the data divided into two groups: brands with 4–10 mg/g of tobacco and those with 11–15 mg/g. Levels of total nicotine in these brands varied

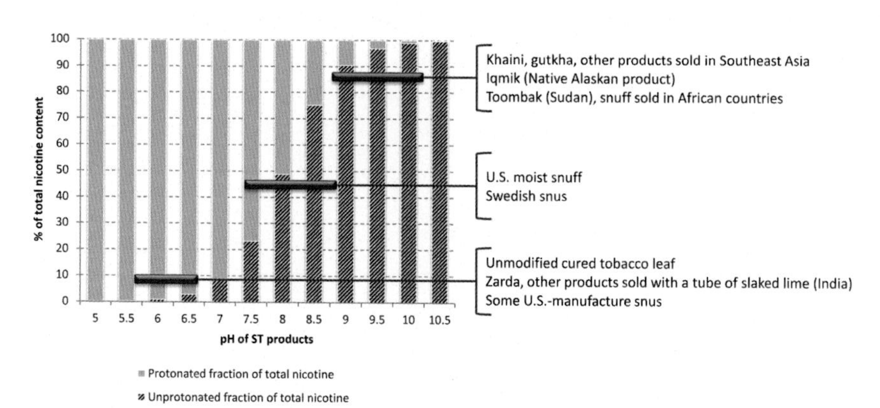

Figure 6.1 Relative contribution of unprotonated and protonated nicotine to the total nicotine content in the range of pH values found in ST products worldwide, with some examples of representative product types for specific pH ranges.[16–19]

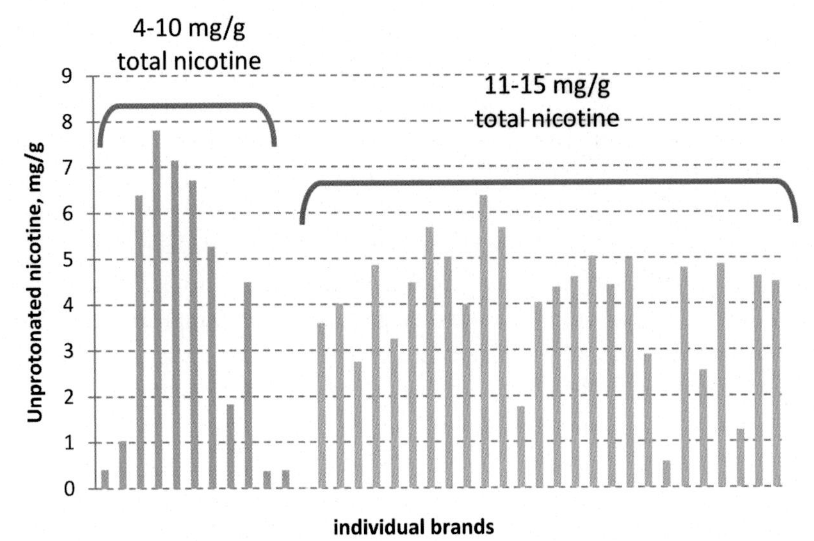

Figure 6.2 Variations of unprotonated nicotine in popular brands of US moist snuff.[20]

3.2-fold, while unprotonated nicotine varied approximately 780-fold. Monitoring of novel US-manufactured ST products, such as snus and dissolvable tobacco, captured variations of unprotonated nicotine levels across different geographic regions where these products were test-marketed, and over time, suggesting that intentional adjustment of pH to appeal to certain consumers may still be done.[18,22,23] In India and other South-East Asian countries, ST is generally used in highly alkaline form, reaching value of pH 10 or even higher.[17,19,24] Products not modified to increase pH are normally sold in these countries together with a small tube of slaked lime so that consumers can mix it in a desired proportion and customize their nicotine intake.[3] This is akin to the "elasticity" of ventilated cigarettes that allows delivery of variable amount of nicotine to smokers depending on smoking intensity.

Tobacco-specific *N*-nitrosamines

TSNA are formed from alkaloids during tobacco curing and processing.[25,26] The resulting yields of TSNA depend on a variety of factors including tobacco type; its cultivation and harvesting practices; levels of nitrate, nitrite, and alkaloids; and processing techniques.[4] Among seven TSNA that have been identified in tobacco products, *N*′-nitrosonornicotine (NNN) and 4-(methylnitrosamino)-1-(3-pyridyl)-1-butanone (NNK) are human carcinogens and are of particular importance as the likely constituents responsible for oral, esophageal, and pancreatic cancer development in ST users.[2,9,27] Levels of these carcinogens vary dramatically across products, as shown in Table 6.1. The highest levels ever reported were in Sudanese toombak: NNN 3080 μg/g and NNK 7870 μg/g of dry product weight.[29] Very high TSNA levels were also reported for some products in India, such as khaini and some brands of zarda.[17,19,24,30] However, the diversity of ST products in India results in coexistence of products with extremely high and extremely low TSNA levels on the same market. For instance, a study that sampled eight types of products in Mumbai found a more than 700-fold variation.[19] TSNA levels in snus sold in Sweden are among the lowest ever reported in commercial ST products: A study of 27 samples found an average NNN level at 0.49 μg/g and NNK at 0.19 μg/g.[28] These low levels are due to the use of pasteurization as the primary tobacco-processing method.[31]

In the United States, a survey of top-selling brands of moist snuff, the most popular ST product type, reported levels of NNN ranged from 2.2 to 42.6 μg/g and NNK from 0.38 to 9.9 μg/g.[20] Another survey focused on products other than moist snuff, such as twist, loose leaf, plug, dry snuff, and some novel products. Variation was even wider: NNN from 0.37 to 31.3 μg/g and NNK from 0.08 to 14.6 μg/g.[32] The lowest levels of TSNA were found in the newer product types such as US-manufactured snus and dissolvable tobacco.[12,23] In addition to variations across product types and brands, differences in NNN and NNK levels within the same brand are commonly observed. For instance, different styles of the Skoal brand showed an approximately 10-fold variation, NNN from 4.5 to 42.5 μg/g and NNK from 0.75 to 9.9 μg/g.[20] A study that focused specifically on the US low-TSNA products also showed considerable variation of NNN and NNK.[23] Differences in NNN and NNK content were found even within brands, depending on the location of purchase.[23,33]

Polycyclic aromatic hydrocarbons

Substantial amounts of PAH are normally formed during the combustion of organic matter. Raw tobacco leaves may contain only trace levels, from environmental pollution. While ST is used without combustion, certain processing and manufacturing practices can lead to contamination of tobacco material with substantial levels of PAH. For example, ST that is made from air-cured tobacco or is pasteurized contains relatively low levels, while products made from fire-cured tobacco, such as US-manufactured moist snuff, contain higher

Table 6.1 Representative variations of NNN and NNK levels across ST product types[a].

Product	Brand or sample	Country	NNN, μg/g	NNK, μg/g	Reference
Betel quid	Handmade	India	0.21	0.03	Stepanov et al.[19]
Chimo	El Tigirito	Venezuela	2.6	1.8	Stanfill et al.[17]
Gul powder	Eagle	Bangladesh	5.2	1.3	Stanfill et al.[17]
Gutkha	RMD	India	0.59	0.24	Stanfill et al.[17]
Khaini	Chaini	India	24.3	2.9	Stepanov et al.[19]
Mawa	Bhola	India	0.59	0.23	Stepanov et al.[19]
Nasvai	Bulk	Uzbekistan	0.64	0.07	Stepanov et al.[28]
Nasvai	Packaged	Kyrgyzstan	1.2	0.19	Stepanov et al.[28]
Naswar	Sample 1	Pakistan	0.36	0.03	Stanfill et al.[17]
Naswar	Sample 2	Pakistan	0.55	0.31	Stanfill et al.[17]
Snuff	Super taxi	South Africa	3.4	0.24	Stanfill et al.[17]
Snuff	Hawken rough wintergreen	United States	3.2	0.76	Richter et al.[20]
Snuff	Grizzly fine cut	United States	8.7	1.7	Richter et al.[20]
Snuff	Red Seal fine cut	United States	14.1	3.9	Richter et al.[20]
Snus	General, Original	Sweden	0.35	0.1	Stanfill et al.[17]
Snus	Camel original	United States	0.79	0.19	Stepanov et al.[12]
Snus	Marlboro rich	United States	1.1	0.23	Stepanov et al.[12]
Toombak	Sample 1	Sudan	119	149	Stanfill et al.[17]
Toombak	Sample 5	Sudan	368	516	Stanfill et al.[17]
Zarda	Hakim pury	Bangladesh	28.6	3.8	Stanfill et al.[17]

[a]Values are per gram of product (wet weight).

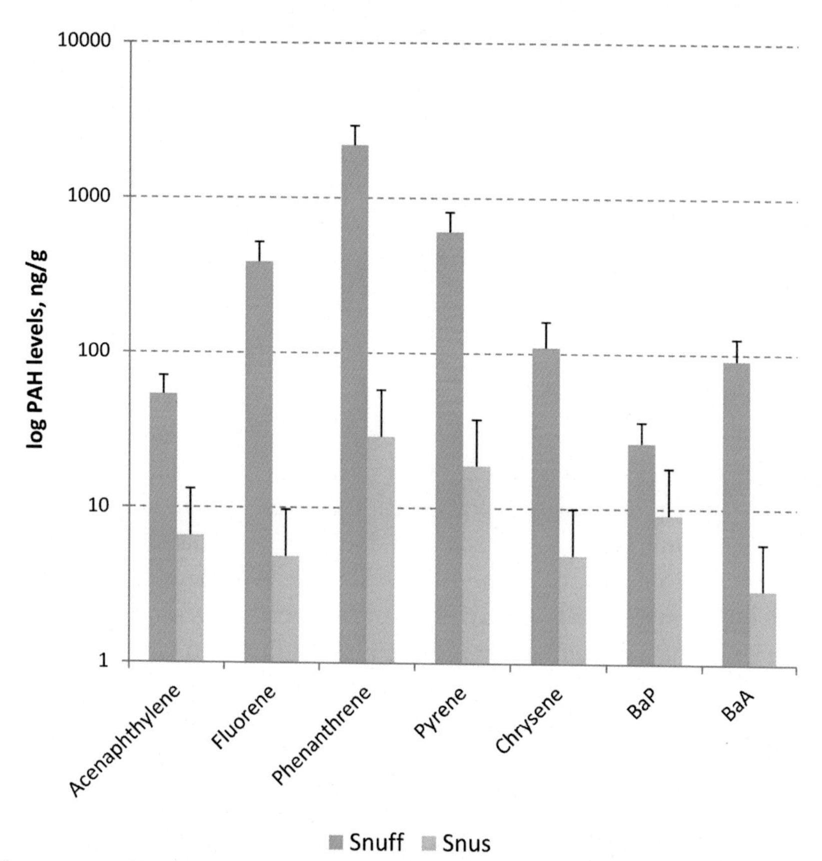

Figure 6.3 Differences in PAH levels in US-manufactured moist snuff (made with fire-cured tobacco) and snus (made without fire-curing).[13] Levels are in logarithmic scale. Bars represent standard deviations for each product type.

levels (Fig. 6.3).[12,13] The sum of 23 PAH in moist snuff ranged from 1250 to 20,200 ng/g (dry weight basis), compared to 901–1940 ng/g in snus.[13] Significant variations in PAH levels across various smokeless products are of concern because many PAH are potent toxicants or carcinogens, including benzo[*a*]pyrene, which is classified as a human carcinogen.[13,34] PAH are also implicated as a cause of cardiovascular disease in smokers.[35]

Volatile aldehydes

Similar to PAH, aldehydes such as acetaldehyde, formaldehyde, and crotonaldehyde can be introduced during fire-curing of tobacco and can also be present in flavoring agents and in parts of industrial equipment and supplies.[36,37] The limited data on levels of aldehydes in ST suggest significant variations.[12,38] A 1992 report showed that formaldehyde ranged from 1.6 to 7.4 μg/g of tobacco, acetaldehyde 1.4–27.4 μg/g, and crotonaldehyde 0.2–2.4 μg/g.[38] A more recent study reported variation by ST type, with products made from pasteurized tobacco such as snus containing somewhat lower levels of formaldehyde and acetaldehyde and higher levels of crotonaldehyde than moist snuff.[12] Given that most aldehydes are irritants and carcinogens, and formaldehyde is a human carcinogen,[37,39,40] elevated levels of these constituents can potentially lead to higher toxicity and carcinogenicity of some ST products.

Metals and metalloids

Levels of metals in soil, and consequently in the tobacco plant, depend on agricultural and geographical factors, such as the use of fertilizers and proximity to metal-processing or metal-utilizing industry.[41,42] In addition, some level of contamination can be expected during manufacture and packaging from nontobacco ingredients and other sources. A review of the limited available reports for ST reveals substantial overall variation for arsenic (0.1–3.5 μg/g), cadmium (0.3–1.88 μg/g), lead (0.23–13 μg/g), and other elements.[8] Limited data also suggest variation across US ST types, with snus containing lower levels of some elements than moist snuff.[4] Some of these elements are human carcinogens.[43]

Measurement of constituent exposure in ST users

Biomarkers of tobacco constituents provide important insights into the impact of constituent variations across products on user exposures. The value of biomarkers is that they account for a range of factors and characteristics that may modify constituent intake by users. Such factors include, but are not limited to, the amount, intensity, and patterns of product use and the rate at which constituents are absorbed and metabolized. Therefore, tobacco constituent biomarkers can serve as an important tool in studies of ST toxicity and carcinogenicity, as well as identifying individuals and populations at risk for disease.

Biomarkers of tobacco constituent exposure

While cotinine, a biomarker of nicotine intake, has been frequently analyzed in a variety of biological matrices, including saliva and blood,[44,45] most other biomarkers are analyzed as urinary metabolites.[46–51] Urine is a robust, noninvasive source of biological material, and metabolites excreted in urine are typically stable end products of a compound's biotransformation. The longer half-life of urinary biomarkers compared to plasma or salivary metabolites is another advantage. Keratin matrices such as hair and toenails are also potentially robust, long-term biomarker sources[52]; however, there is a lack of validated toenail assays, and the amount of sample available is limited.

Fig. 6.4 shows the structures of some key urinary biomarkers of tobacco constituents. These biomarkers have been analytically validated and applied in numerous studies of tobacco users. It should be noted that most of the studies that employed these biomarkers dealt with cigarette smokers rather than ST users. Nonetheless, the information obtained is helpful in assessing their utility for investigating exposures in ST users. For instance, studies in smokers and nonsmokers provided valuable insights into the specificity of some biomarkers to tobacco exposures (e.g., for nicotine and TSNA) or sensitivity of other biomarkers to differences in the level of exposure (e.g., for PAH and aldehydes).[46] Nicotine intake is measured by analyzing urinary total cotinine or urinary total nicotine equivalents (TNE), which is the sum of urinary nicotine, cotinine, and 3′-hydroxycotinine, plus their glucuronides (Fig. 6.4). Approximately 73%–96% of the nicotine dose is excreted as TNE, making it an excellent biomarker of nicotine exposure, superior to total cotinine.[53,54] Numerous studies of carcinogen and toxicant uptake in smokers demonstrated that both the total cotinine and TNE in urine correlate with biomarkers of other key tobacco-related toxicants and carcinogens.[50,55–62] In addition, the ratio between 3′-hydroxycotinine and cotinine is being used as an indicator of nicotine metabolism rate. Studies in smokers show

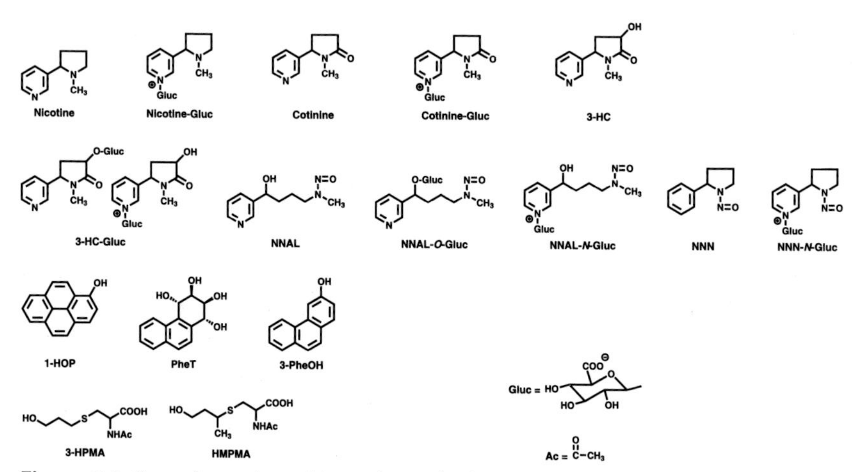

Figure 6.4 Some key urinary biomarkers of tobacco constituents measured in users.

that, compared to fast nicotine metabolizers, slower metabolizers smoke fewer cigarettes, which would play a role in the extent of toxicant exposure.[63,64] For example, compared to Caucasians, African-Americans have slower metabolism of nicotine and tend to smoke fewer cigarettes; however, they take in more nicotine per cigarette (potentially due to more intense puffing) and experience a higher incidence of smoking-related disease.[65–67]

Exposure to NNN and NNK is measured by urinary total NNN and total 4-(methylnitrosamino)-1-(3-pyridyl)-1-butanol (NNAL), respectively, with both biomarkers being the sum of the corresponding metabolite and its glucuronide(s).[46,68,69] Many studies have reported levels of total NNN and NNAL in cigarette smokers and other tobacco product users[55,58,61,62,70]; these biomarkers are not found in urine of nontobacco users unless they have been exposed to secondhand smoke.[71] In the case of ST users, urinary total NNN is more relevant than total NNAL because NNN is an oral and esophageal carcinogen while NNK causes lung cancer (which is not prevalent in ST users).[2,9,72] However, measurement of total NNAL in urine has the advantage of being a longer-term surrogate biomarker of TSNA exposure; its elimination half-life extends more than 2 weeks, while approximately 90% of urinary total NNN is cleared in 3 days after smoking cessation.[73–75] On the other hand, the shorter half-life of urinary total NNN makes it a more suitable biomarker for short-term studies.

Examples of biomarkers that are related, but not specific, to tobacco exposure include urinary 1-hydroxypyrene (1-HOP), phenanthrene tetraol (PheT), and a panel of mercapturic acids. 1-HOP and PheT are metabolites of pyrene and phenanthrene, respectively; both are noncarcinogenic PAH. Pyrene and phenanthrene are always present in PAH mixtures, but their levels are much higher than those of key carcinogenic PAH such as benzo[*a*]pyrene. 1-HOP has a longer history of applications in human exposure studies and is widely accepted as a biomarker of PAH exposure.[76] However, PheT has an advantage in that it also incorporates a major pathway of metabolic activation via diol epoxide metabolites, which is similar to cytochrome P450 1A1- and 1B1-catalyzed metabolic activation of benzo[*a*]pyrene.[77–79] Levels of PheT in human urine correlate with those of 1-HOP and with benzo[*a*]pyrene tetraol, a metabolite of benzo[*a*]pyrene.[76] While urinary PheT can serve as a biomarker of PAH metabolic activation,[80] another phenanthrene metabolite excreted in urine, 3-hydroxyphenanthrene (3-PheOH), represents PAH exposure and detoxification.[81] Therefore, the ratio of PheT to 3-PheOH can potentially be used as an indicator of a tobacco user's capacity to metabolically activate PAH. Urinary mercapturic acids

SPMA, 3-HPMA, and HMPMA are biomarkers of benzene, acrolein, and crotonaldehyde, respectively.[46] These metabolites are formed by conjugation with glutathione to the α,β-unsaturated carbonyl system, followed by reduction of the carbonyl group and normal cellular processing of the glutathione conjugates to *N*-acetyl cysteine conjugates, which are excreted in the urine. Given that these biomarkers are not specific to tobacco, their levels can be confounded by exposure to environmental pollution.

Biomarker studies in ST users

Biomarker-based studies in ST users are relatively scarce. Clinical studies have shown that blood plasma nicotine levels in users of higher pH products are higher than nicotine levels delivered by lower pH products.[82] Studies of NNN and NNK exposure in the United States showed that users of moist snuff take in these carcinogens at levels similar to, or higher than, cigarette smokers.[69,83] Use of the Alaskan Native ST product iqmik was associated with high nicotine exposure due to mixing ash, an alkaline agent that increases the pH of the product, with tobacco leaves or commercial ST, but lower NNK exposure compared to exclusively commercial ST users.[84] Some studies have examined the relationship between biomarkers of exposure and patterns of ST use. For instance, regression analysis in a study of 54 users showed that daily duration of use (portion or "dip" duration) was significantly associated with total cotinine and total NNAL.[85] Urinary total nicotine and total NNAL levels were also associated with tins of product used per week.[85,86] Urinary cotinine levels were also related to frequency, durational measures, and weight of dip.[87–90] However, most of these studies did not take into account constituent levels in the products.

Studies investigating the effects of constituent level variations across ST products on exposures in users are even more limited and are mostly focused on nicotine and TSNA. A cross-sectional comparison of biomarker results from studies that examined exposure to NNN and NNK from various ST brands showed that urinary total NNAL in ST users is strongly associated with the level of NNK in the product.[91] A more recent study directly investigated the relationship between TSNA levels in ST and urinary biomarkers by targeted recruiting of users of specific ST brands that differed in TSNA levels.[92] Results of univariate analyses showed significant correlations between levels of NNK in the product with total NNAL in urine of 359 ST users, and of NNN in the product and total NNN in urine. These associations remained significant in the subsequent multiple regression analyses that included the amount and intensity of product use. Another approach to investigating the effect of constituent levels in an ST product on the relevant exposures in users is to conduct a study in which ST users switch between products that differ substantially in their constituent levels. One such study switched users of US moist snuff to low-TSNA General Snus or nicotine patch for 4 weeks.[93] The results showed that switching led to a significant reduction in urinary total NNAL, while no differences were observed in the amount of product used or urinary cotinine levels, indicating that the reduction in urinary total NNAL was not due to decreased product use. Together, these studies provide evidence that levels of harmful constituents in ST are strong independent predictors of exposure to these constituents in ST users.

Relationship between tobacco constituent biomarkers and cancer risk

In addition to serving as reliable measures of harmful constituent intake, some biomarkers have been shown to be associated in a dose-dependent manner with cancer risk in smokers. A case—control study of the dose—response relationship between urinary total NNN and esophageal cancer risk, based on the Shanghai Cohort Study, analyzed prospectively collected urine samples from 77 smokers who developed esophageal cancer and 223 matched cancer-free smokers.[94] The results showed that, after adjustment for confounders such as smoking history and intensity, odds ratio of esophageal cancer risk in the highest tertile of urinary total NNN compared to the lowest tertile was 17.0 (95% confidence interval 4.0 to 72.8). Analysis of urinary total NNAL and PheT from lung cancer cases and matched controls in the same cohort showed that both biomarkers were independently associated with lung cancer risk.[95] Similar predictive power of total NNAL for lung cancer risk in smokers was demonstrated in two other cohorts.[96,97] Together, these studies indicate that biomarker-assessed levels of TSNA and PAH intake predict cancer risk in smokers, and the relationship is also likely to be true for ST users. For instance, urinary total NNAL in ST users was associated dose-dependently with the presence and degree of oral leukoplakia, which is a premalignant condition for oral cancer.[98]

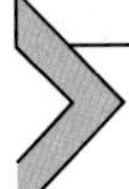

Analytical methodologies for the analysis of tobacco constituents and biomarkers

Given the wide range of key harmful constituent levels across ST products and the important role of biomarker measurements for understanding exposures and disease risk, the use of accurate and robust analytical methods is imperative. Appropriate quality control measures are also necessary to ensure accuracy and precision of results. Following are examples of analytical methodologies used for the analysis of constituents and biomarkers discussed in this chapter.

Tobacco constituent analysis

General considerations for tobacco sample handling

Avoiding laboratory contamination: Some laboratories are engaged in research projects that require analyses of both the constituents in tobacco products and biomarkers of exposure to these constituents in biological samples. For some such measurements, for instance, nicotine and NNN analyses, the difference in the levels of the same chemical compound between the two matrices (e.g., tobacco and urine) may be several orders of magnitude. Therefore, preventing laboratory contamination with tobacco constituents and avoiding the potential cross-contamination of biological samples is a critical element in the tobacco analytical research. There are several general rules that can be followed to minimize chances for contamination. These include using dedicated laboratory space and a designated fume hood for tobacco sample preparation, and using only designated laboratory equipment and supplies such as pipettors, vial racks, analytical balances, pH-meters, SpeedVac vacuum concentrators, and other supplies for tobacco product analyses.

Preserving the product for future analyses: Typically, only a small amount of tobacco material is necessary for carrying out the analyses of most constituents discussed in this chapter. Therefore, the same tin or packet of ST can be used for multiple assays, still leaving enough for future analyses of additional constituents. Certain omissions can interfere with reuse or the accuracy of future measurements. Examples include failure to properly label the product sample, contamination, moisture loss, and product aging. Each tin or packet should be assigned a unique ID, and the label should be placed on the container in a way that does not cover any important information on the original manufacturer's label, such as brand and variety name, product weight, and expiration date. To take a sample for weighing, use disposable forceps to avoid cross-contamination from a previously weighed sample. If too much material was taken out, the excess should not be returned to the container. While working with moist snuff and other products with high moisture content, moisture loss can occur rapidly after the container is opened. Significant moisture loss will affect the calculated levels of constituents per gram of product; therefore, a sample for moisture content determination should be taken out first, immediately after opening the container. To minimize product aging and any significant changes in chemical composition and/or moisture content, the container should be sealed and refrigerated or frozen for prolonged storage.

Analysis of nicotine and other tobacco alkaloids

Nicotine can be analyzed by gas chromatography (GC)-mass spectrometry (MS).[12] Liquid chromatography (LC)-MS/MS can also be used.[99,100] The convenience of LC-MS/MS consists in its ability to simultaneously analyze the minor tobacco alkaloids nornicotine, anatabine, and anabasine in one procedure. Analysis of minor tobacco alkaloids in ST users can be informative because some studies suggest that nornicotine and anatabine may contribute to the addiction potential of ST. Also, analysis of anatabine and anabasine in urine of study participants is sometimes used to verify abstinence from tobacco products. Lastly, endogenous nitrosation of nornicotine in the oral cavity, the stomach, or elsewhere in the body can result in the formation of NNN, adding to carcinogen exposure.[101] Therefore, measurement of these constituents in ST products can aid in interpretation of study results and characterization of carcinogenic potential of a particular product. The advantage of LC-MS/MS for this purpose is that nornicotine can be analyzed directly, while GC-MS would require converting it to a tertiary amine derivative via reductive alkylation with propionaldehyde and sodium borohydride.[12] For LC-MS/MS analysis, the sample is extracted with 10 mM ammonium acetate containing 5% methanol, and the extract is mixed with stable-isotope-labeled internal standards (e.g., [D_3]nicotine, [D_4]nornicotine, [D_4]anabasine, and [D_4]anatabine) and subjected to serial dilution. The prepared samples are analyzed by LC-MS/MS on a Hypercarb column (Thermo Scientific) using 10 mM ammonium acetate (with 0.01% formic acid) and methanol as the mobile phase.[43,44,102]

Analysis of tobacco-specific N*-nitrosamines*

Five TSNA are commonly reported for ST products: NNN, NNK, NNAL (which is not only an NNK biomarker in vivo but can also be present in small amounts in tobacco), *N′*-nitrosoanatabine (NAT), and *N′*-nitrosoanabasine (NAB). Among these, NNN, NNK, and NNAL are of primary interest because of their carcinogenicity.[9] However, NAT and NAB can be measured simultaneously in the same assay, and their levels can provide important insights into the tobacco type and the overall carcinogenic potential of the product. Previous routine analyses of TSNA often used GC coupled with the nitrosamine-selective Thermal Energy Analyzer.[12] However, LC-MS/MS has much higher sensitivity and selectivity and allows a less laborious and streamlined sample preparation.[103,104] Briefly, the sample is extracted with 10 mM ammonium acetate, and the extract is mixed with stable-isotope-labeled internal standards (e.g., $^{13}C_6$-labeled NNN, NNK, and NNAL) and subjected to liquid–liquid extraction on ChemElut or similar cartridges prior to analysis by LC-MS/MS. The LC-MS/MS analysis[104] is done in the positive ion electrospray mode with selected reaction monitoring for m/z 178 → 148 for NNN, m/z 208 → 178 for NNK, m/z 210 → 93 for NNAL, m/z 190 → 160 for NAT, m/z 192 → 162 for NAB, and corresponding transitions for internal standards.

Analysis of polycyclic aromatic hydrocarbons

PAH can be analyzed by the GC-MS method originally developed by Ding et al.[105,106] and modified by Stepanov et al. for ST analysis, allowing simultaneous determination of at least 23 different PAH.[13] The method can be modified to target a smaller number of PAH of interest; for instance, it can be limited to benzo[*a*]pyrene (the representative carcinogenic PAH) and pyrene and phenanthrene (sources of biomarkers 1-HOP and PheT, respectively). For PAH analysis, the sample is spiked with a mix of stable-isotope-labeled internal standards, extracted with cyclohexane, and purified on normal-phase, silica-based cartridges. Analysis is carried out by GC-MS in the positive electrospray ionization mode, monitoring for ions of interest (Table 6.2).[13]

Analysis of aldehydes

There are not many reports on aldehyde assays in ST products. However, methods used for cigarette smoke and other samples exist and can be adapted. Aldehydes can be extracted with methylene chloride and derivatized by mixing with acidified 2,4-dinitrophenylhydrazine (DNPH) solution.[107,108] The DNPH derivatives can be analyzed by GC-MS[12] and high-performance liquid chromatography (HPLC) with UV detection,[107–109] or ultrahigh-pressure LC-MS/MS (UPLC-MS/MS).[110]

Analysis of metals and metalloids

For the elemental analysis, ST needs to be digested in nitric acid or a mix of nitric acid and hydrogen peroxide.[8] The digested sample can be analyzed by either atomic absorption spectroscopy or inductively coupled plasma (ICP) MS. A streamlined and optimized method for metal analysis in cigarette

Table 6.2 Quantitation and confirmation ions monitored in the analysis of PAH by GC-MS[13].

Analyte	Quantitation ion (*m/z*)	Confirmation ion (*m/z*)
Naphthalene	128.2	127.2
Acenaphthylene	152.1	151.1
Acenaphthene	154.2	152.2
Fluorene	166.2	165.2
Phenanthrene	178.2	176.2
Anthracene	178.2	176.2
Fluoranthene	202.3	200.3
Pyrene	202.3	200.3
Benz[*a*]anthracene	228.3	226.3
Chrysene	228.3	226.3
1-Methylchrysene	242.3	239.3
3-Methylchrysene	242.3	239.3
4-Methylchrysene + 6-methylchrysene	242.3	239.3
5-Methylchrysene	242.3	239.3
Benzo[*b*]fluoranthene + Benzo[*j*] fluoranthene	252.3	250.3
Benzo[*k*]fluoranthene	252.3	250.3
Benzo[*e*]pyrene	252.3	250.3
Benzo[*a*]pyrene	252.3	250.3
Indeno[1,2,3-*cd*]pyrene	276.3	274.3
Dibenz[*a,h*]anthracene	278.3	276.3
Benzo[*g,h,i*]perylene	276.3	274.3

tobacco was recently reported by Fresquez et al.[111] and can be applied to ST. The method uses microwave digestion and ICP-MS for a range of elements but not for mercury, which is analyzed by a direct combustion analyzer.

Measurement of pH and moisture content

Given that pH determines the proportion of nicotine present in the biologically available unprotonated form, measurement of pH should always be included and reported together with nicotine content. To measure pH, an aqueous extract of tobacco is prepared at the approximate ratio of 1 mL ultrapure H_2O per 100 mg of sample.[12,19] To facilitate extraction, the sample can be placed in a sonicator for 5–10 min followed by passive extraction at room temperature for an additional 20–30 min. Alternatively, tubes with sample-water mix can be placed in a benchtop shaker for the whole duration of the extraction. Following extraction, the sample is centrifuged to pellet down the tobacco particles, and the pH of the aqueous extract is measured with a pH meter.

Moisture content of different ST products can vary from 5% to 60%. Therefore, it is important to express the amounts of various toxicants and carcinogens per gram dry weight of product to aid in interpretation of the results, especially when the amounts of constituents are compared across products or by different laboratories. The easiest approach is a gravimetric method, in which moisture is measured by calculating the difference in the weight of the sample before and after it was dried for 3 h in a heating block set at 99°C.

Standardized WHO TobLabNet methods for the analysis of tobacco constituents

The WHO Network of Tobacco Laboratories (TobLabNet) has been developing standardized methods for tobacco constituent analyses with the goal of advancing tobacco control efforts through enabling the implementation of Articles 9 and 10 of the WHO Framework Convention on Tobacco Control (FCTC). Such standardized methods are needed to ensure cross-reference, reliability, and credibility of constituent data generated by laboratories in different countries. The initial method and standard operating procedure (SOP) development was focused primarily on cigarettes as the most popular tobacco product in the majority of countries that are members of the FCTC. Currently available TobLabNet SOPs are listed in Table 6.3. Most SOPs developed for cigarette smoke analysis can be adapted to ST. Some (e.g., SOP 04, SOP 06, and SOP 07) were developed for cigarette tobacco and can be directly applied to ST. Therefore, whenever possible, chemical analyses of ST should be conducted according to WHO TobLabNet SOPs to ensure accurate, reproducible results that can serve as a basis for regulatory measures. Note that the TobLabNet SOPs are consistent with most assays discussed in this chapter. The significance of SOPs is that they outline detailed step-by-step procedures and account for major factors that may influence the outcome of measurements.

Urinary biomarker analysis

General considerations for urine sample handling

Sample integrity: While some tobacco constituent biomarkers, such as TNE and NNAL, are fairly stable, others may be sensitive to sample storage conditions. For instance, NNN can be formed in urine after collection if the urine contains nitrite and nornicotine. Furthermore, analyses of additional

Table 6.3 Available WHO TobLabNet standard operating procedures for the analysis of priority constituents in tobacco products.

Constituent or measure	SOP (year developed)
Intense smoking regimen	SOP 01: SOP for intense smoking of cigarettes (2012)
Method validation for tobacco product analysis	SOP 02: Validation of analytical methods of tobacco product contents and emissions (2017)
NNK and NNN in cigarette smoke	SOP 03: Determination of tobacco-specific nitrosamines in mainstream cigarette smoke under ISO and intense smoking conditions (2014)
Nicotine in cigarette tobacco	SOP 04: Determination of nicotine in cigarette tobacco filler (2014)
BaP in cigarette smoke	SOP 05: Determination of benzo[*a*]pyrene in mainstream cigarette smoke (2015)
Propylene glycol, glycerol, and triethylene glycol in cigarette tobacco	SOP 06: Determination of humectants in cigarette tobacco filler (2015)
Ammonia in cigarette tobacco	SOP 07: Determination of ammonia in cigarette tobacco filler (2016)
Acetaldehyde, acrolein, and formaldehyde in cigarette smoke	SOP 08: Determination of aldehydes in mainstream cigarette smoke under ISO and intense smoking conditions (2018)
1,3-Butadiene and benzene in cigarette smoke	SOP 09: Determination of volatile organics in mainstream cigarette smoke under ISO and intense smoking conditions (2018)
Nicotine and carbon monoxide in cigarette smoke	SOP 10: Determination of nicotine and carbon monoxide in mainstream cigarette smoke under intense smoking conditions (2016)

biomarkers sometimes become important after the analysis of the target set is completed, and the additional biomarkers could be sensitive to sample storage conditions or freeze—thaw cycles. Therefore, preparing small sample aliquots and promptly freezing them after urine is collected is an important approach to preserving sample integrity. Aliquots can be thawed on the benchtop at room temperature or in a water bath at 37°C before analysis.

Avoiding sample contamination with tobacco constituents: Similar to the best practices outlined for tobacco sample handling, urine and other biological samples that are being analyzed for trace levels of biomarkers should be processed in a designated laboratory space and fume hood, using laboratory equipment and supplies that are not being used for the analyses of tobacco products or other potential sources of contamination with tobacco constituents.

Analysis of total nicotine equivalents

For MS-based assays, only a small volume of urine is required for TNE analysis.[53] Ten microliter diluted 1:10 is usually sufficient. However, if the urine is collected from an occasional user or a user of a low-nicotine product, it may be diluted less or processed undiluted. Given the sensitivity and robustness of nicotine metabolite assays, sample processing for TNE can be done in a 96-well plate format.[53] Briefly, stable-isotope-labeled internal standards (e.g., $[CD_3]$nicotine, $[CD_3]$cotinine, and $[CD_3]3'$-hydroxycotinine) are added to the plates which are treated with β-glucuronidase to release free metabolites from their glucuronide conjugates. Samples are then subjected to reversed-phase, solid-phase extraction, and the eluents are analyzed by LC-MS/MS with transitions m/z 163 → m/z 130 for nicotine, m/z 177 → m/z 98 for cotinine, m/z 193 → m/z 134 for 3′-hydroxycotinine, and corresponding transitions for the internal standards.

Analysis of urinary biomarkers of tobacco-specific N*-nitrosamines*

Several variations of the LC-MS/MS method for total NNN and total NNAL have been published, and most of them are robust and analytically validated. The assays for both biomarkers share similar processing steps, but there are important differences. Urinary total NNN accounts for only about 1%–3% of NNN dose[75] and therefore is normally present at much lower levels than total NNAL. Also, NNN is easily formed from nornicotine and nitrite, and urine of tobacco users always contains nornicotine as a metabolite of nicotine.[54] Lastly, NNN tends to be more susceptible to ion suppression and therefore requires a more thorough sample purification prior to LC-MS/MS analysis. Following are the procedures for each biomarker.

Urinary total NNN: Before purification, samples are mixed with a stable-isotope-labeled standard (e.g., $[^{13}C_6]$NNN) and treated with β-glucuronidase to convert any NNN-*N*-glucuronide to free NNN.[112] The sample is then purified sequentially using supported liquid–liquid extraction cartridges; mixed-mode, reversed-phase, and cation-exchange cartridges; and normal-phase, solid-phase extraction cartridges. To prevent artifactual formation of NNN, ascorbic acid or another nitrite-trapping agent should be added during the liquid—liquid and mixed-mode extraction, when nornicotine and nitrite are present.[113] The purified sample is analyzed by LC-MS/MS, monitoring for m/z 178 → 148 for NNN and the corresponding transition for the internal standard. A more sensitive LC-nanoelectrospray ionization-high-resolution MS/MS method was also reported in a study that analyzed the urine of e-cigarette users who are exposed to very low levels of NNN.[114] The method was developed for an Orbitrap Fusion Tribrid mass spectrometer operated in the product ion scan mode, using a hand-packed capillary column with Luna C_{18}—bonded separation media. The analysis was performed using accurate mass extracted ion chromatograms of m/z 148.0995 $[C_9H_{12}N_2]^+$ (parent ion m/z 178.1) for NNN and corresponding fragment $[^{13}C_6]$NNN internal standard, with a mass tolerance of 5 ppm. This method can be applied for the analysis of NNN in users of low-TSNA ST.

Urinary total NNAL: The analysis requires fewer purification steps than for NNN and can be done using 96-well plate technology.[115] Samples are mixed with stable-isotope-labeled internal standard, treated with β-glucuronidase to convert NNAL glucuronides to free NNAL, and enriched by supported liquid extraction plate and mixed-mode, reverse-phase, cation-exchange, and solid-phase extraction. The purified sample is analyzed by LC-MS/MS with selected reaction monitoring m/z 210.12 → m/z 93.14 for NNAL and corresponding transition for the internal standard. A method has also been reported for combined analysis of urinary total NNAL and total cotinine.[116] It allows simultaneous analysis of NNAL and cotinine by monitoring m/z 178 → 98 for cotinine, where m/z 178 is the $[M + H]^+$ ion resulting from the naturally occurring $[^{13}C]$cotinine.

Analysis of urinary biomarkers of polycyclic aromatic hydrocarbons

The choice between analyzing urinary 1-HOP or PheT as a biomarker of PAH exposure will depend on whether assessing PAH metabolism is important for the study goals, and on the availability of the required analytical equipment.

Urinary 1-HOP: The sample is treated with β-glucuronidase, mixed with internal standard, purified by reversed-phase, solid-phase extraction, and analyzed by HPLC with fluorescent detection.[117,118] A unique feature of this assay is that a stable-isotope-labeled internal standard can be used even though the analysis does not involve MS detection. This is because the internal standard, $[D_9]$1-HOP, contains a sufficient number of deuterium atoms to achieve a near-baseline chromatographic separation from the unlabeled biomarker.

Urinary PheT and 3-PheOH: Analysis of phenanthrene biomarkers is based on a sensitive and selective MS technique and therefore can be done by 96-well plate technology.[115] For PheT, urine is mixed with [$^{13}C_6$]PheT or [D_{10}]PheT internal standard, treated overnight with β-glucuronidase and aryl sulfatase, purified on styrene-divinylbenzene sorbent, and silylated with bis-trimethylsilyltrifluoroacetamide. The formed derivative PheT-tetratrimethylsilyl ether is then analyzed by GC-negative ion chemical ionization MS with selected ion monitoring at m/z 372 for the analyte and corresponding transition for the internal standard. A similar procedure except for the derivatization is used for the analysis of 3-PheOH, monitoring m/z 266 for 3-PheOH and corresponding transition for its internal standard.

Analysis of urinary mercapturic acids

A high-throughput method for mercapturic acid analysis in 96-well plate format has been developed and analytically validated.[119] The samples are added to the plates along with the corresponding internal standards, and solid-phase extraction is done on Oasis MAX mixed-mode, reverse-phase, and anion-exchange plates. The purified samples are then analyzed by LC-MS/MS in the atmospheric pressure chemical ionization mode. For the mercapturic acids discussed earlier in this chapter, the transitions monitored are m/z 238 → 109 for SPMA, m/z 220 → 91 for 3-HPMA, and m/z 234 → 105 for MHPMA.

Analysis of metals in urine

Metals and metalloids in urine are analyzed by methods similar to tobacco analyses by ICP-MS. Prior to analysis, urine is diluted with nitric acid, and Triton X-100 can be added as a surfactant.[120–122]

Quality control considerations

Analytical assays for tobacco constituents and biomarkers should follow quality control procedures consistent with US Centers for Disease Control and Prevention and US Food and Drug Administration (FDA) guidance. Positive and negative controls must be included in each set of samples in order to monitor for analytical variation and potential contamination, respectively, as a way to monitor the performance and accuracy of each assay. For tobacco product analysis, reference ST products such as CRP1 and CRP2 should be used as positive controls. These products are manufactured by the Center for Tobacco Reference Products at the University of Kentucky Tobacco Research and Development Center, with the sole purpose of being used for quality control (to monitor accuracy and repeatability of ST product analyses) and method validation. CRP1 is manufactured to be representative of snus in its composition and levels of key tobacco constituents, and CRP2 is representative of US moist snuff. Newer versions may become available, along with reference products for other ST types, and should be acquired and used in ST product analyses. For urine samples, pooled smokers' urine with established levels of biomarkers can be used. Water blanks or extraction solutions are used in specific protocols as negative controls. The controls should be processed together with the rest of the samples, using identical procedures and internal standards. While the methods described above have been analytically validated with respect to linearity, accuracy, and precision, the analysis of each batch of samples has to start with the injection of calibration standards, with the calibration curves covering the entire range of values that are found in samples. The relative standard deviation of the quality control samples must be less than 10% for intraday and less than 15% for interday assays.

Policy implications

Standards for carcinogen and toxicant levels in ST products have been proposed by the WHO Study Group for Tobacco Product Regulation (TobReg) and the FDA. Evidence to support this proposal is demonstrated by the variability of harmful and potentially harmful constituents across ST products, the relationship between these constituent levels and biomarkers of exposure, and the differences in health effects depending on the extent of carcinogens and toxicants in ST used by a country (e.g., Swedish snus vs. products in India), as discussed in this and other chapters. Furthermore, there is clear evidence that these harmful constituents can be reduced by using certain types of tobacco and by incorporating specific methods for growing, processing, curing, and storing, as discussed in Chapter 8. The WHO TobReg through Article 9 of FCTC recommended that a governmental regulatory agency might issue a mandate to impose an upper limit to the toxicants in ST products.[123] For industry-manufactured products, the recommended limit for NNN plus NNK was 2 μg/g of tobacco dry weight and for benzo[*a*]pyrene 5 ng/g. Furthermore, WHO TobReg recommended that levels of arsenic, cadmium, and lead should be monitored. The FDA issued an Advanced Notice of Proposed Rulemaking on January 22, 2017, to limit the level of NNN to 1 μg/g of tobacco dry weight in finished ST products. The FDA estimated that about 12,700 new cases of oral cancer and 2200 oral cancer deaths would be averted in the United States in the 20 years following the implementation of the proposed product standard. Additionally, the FDA stated that the rule would require finished ST products to have an expiration date based on the demonstration that the product still adheres to the NNN product standard and specification for the method of storage (e.g., refrigeration or room temperature). Swedish Match for years has implemented standards for their ST products (GothiaTek standard, https://www.swedishmatch.com/Snus-and-health/GOTHIATEK/GOTHIATEK-standard/), demonstrating the feasibility of mandating limits of harmful constituents in ST.

Although imposing product standards might lead to fewer ST-related deaths, the ultimate goal is to prevent initiation of ST use and to help people quit. Two methods that could facilitate reduction in prevalence of ST use are to regulate free nicotine or pH to reduce addictiveness, and to ban flavorants and sweeteners that make the product appealing (WHO TobReg Series 989). Such regulations would likely result in a great deal of controversy. Some tobacco control stakeholders argue that ST is less harmful than cigarettes, and to facilitate switching from cigarettes to ST, the product must have pharmacokinetics similar to cigarettes and must be appealing. Other tobacco control stakeholders believe that ST is not harmless, and any advocacy for it would result in significant uptake among youth and addiction of converted smokers to another product. Regardless of this controversy, it is still important to consider government regulations to reduce the toxicity of ST, or at a minimum to educate the public about the toxicity and the variability of harmful constituents in ST products so that they can make an informed decision about use.

References

1. Roberts NL. Natural tobacco flavor. *Rec Adv Tob Sci*. 1988;37:49–81.
2. IARC. *Smokeless Tobacco and Tobacco-specific Nitrosamines. IARC Monographs on the Evaluation of Carcinogenic Risks to Humans*. Vol. 89. Lyon, FR: International Agency for Research on Cancer; 2007.
3. Stanfill SB, Stepanov I. Chapter 3: a global view of smokeless tobacco products. In: U.S. Department of Health and Human Services CfDCaPaNIoH, National Cancer Institute, ed. *Smokeless Tobacco and Public Health: A Global Perspective*. Vol NIH Publication No. 14-79832014.
4. Stepanov I, Hatsukami D. Call to establish constituent standards for smokeless tobacco products. *Tob Regul Sci*. 2016;2:9–30.
5. Burton HR, Dye NK, Bush LP. Distribution of tobacco constituents in tobacco leaf tissue. 1. Tobacco-specific nitrosamines, nitrate, nitrite, and alkaloids. *J Agric Food Chem*. 1992;40:1050–1055.

6. Chamberlain WJ, Chortyk OT. Effects of curing and fertilization on nitrosamine formation in bright and burley tobacco. *Beitr Tabakforsch Int.* 1992;15:87–92.
7. DeRoton C, Wiernik A, Wahlberg I, Vidal B. Factors influencing the formation of tobacco-specific nitrosamines in French air-cured tobaccos in trials and at the farm level. *Beitr Tabakforsch Int.* 2005;21:305–320.
8. Pappas RS, Stanfill SB, Watson CH, Ashley DL. Analysis of toxic metals in commercial moist snuff and Alaskan iqmik. *J Anal Toxicol.* 2008;32(4):281–291.
9. Hecht SS. Biochemistry, biology, and carcinogenicity of tobacco-specific *N*-nitrosamines. *Chem Res Toxicol.* 1998;11:559–603.
10. Fischer S, Spiegelhalder B, Preussmann R. Preformed tobacco-specific nitrosamines in tobacco - role of nitrate and influence of tobacco type. *Carcinogenesis.* 1989;10: 1511–1517.
11. Andersen RA, Burton HR, Fleming PD, Hamilton-Kemp TR. Effect of storage conditions on nitrosated, acylated, and oxidized pyridine alkaloid derivatives in smokeless tobacco products. *Cancer Res.* 1989;49:5895–5900.
12. Stepanov I, Jensen J, Hatsukami D, Hecht SS. New and traditional smokeless tobacco: comparison of toxicant and carcinogen levels. *Nicotine Tob Res.* 2008;10(12): 1773–1782.
13. Stepanov I, Villalta PW, Knezevich A, Jensen J, Hatsukami DK, Hecht SS. Analysis of 23 polycyclic aromatic hydrocarbons in smokeless tobacco by gas chromatography-mass spectrometry. *Chem Res Toxicol.* 2010;23:66–73.
14. Tomar SL, Henningfield JE. Review of the evidence that pH is a determinant of nicotine dosage from oral use of smokeless tobacco. *Tob Control.* 1997;6:219–225.
15. Henningfield JE, Fant RV, Buchhalter AR, Stitzer ML. Pharmacotherapy for nicotine dependence. *CA Cancer J Clin.* 2005;55(5):281–299.
16. Richter P, Spierto FW. Surveillance of smokeless tobacco nicotine, pH, moisture, and unprotonated nicotine content. *Nicotine Tob Res.* 2003;5(6):885–889.
17. Stanfill SB, Connoly GN, Zhang L, et al. Global surveillance of oral tobacco products: total nicotine, unionized nicotine and tobacco-specific *N*-nitrosamines. *Tob Control.* 2011;20(3):e2. https://doi.org/10.1136/tc.2010.037465.
18. Stepanov I, Biener L, Yershova K, et al. Monitoring tobacco-specific N-nitrosamines and nicotine in novel smokeless tobacco products: Findings from Round 2 of the New Product Watch. *Nicot Tob Res.* 2014. https://doi.org/10.1093/ntr/ntu026. Advance access.
19. Stepanov I, Gupta PC, Parascandola M, et al. Constituent variations in smokeless tobacco purchased in Mumbai, India. *Tob Regul Sci.* 2017;3(3):305–314.
20. Richter P, Hodge K, Stanfill S, Zhang L, Watson C. Surveillance of moist snuff: total nicotine, moisture, pH, un-ionized nicotine, and tobacco-specific nitrosamines. *Nicotine Tob Res.* 2008;10(11):1645–1652.
21. Connolly GN. The marketing of nicotine addiction by one oral snuff manufacturer. *Tob Control.* 1995;4:73–79.
22. Stepanov I, Jensen J, Biener L, Bliss RL, Hecht SS, Hatsukami DK. Increased pouch sizes and resulting changes in the amounts of nicotine and tobacc-specific N-nitrosamines in single pouches of Camel Snus and Marlboro Snus. *Nicotine & Tobacco Research.* 2012. https://doi.org/10.1093/ntr/ntr292. Advance Access, publishe January 17, 2012.
23. Stepanov I, Biener L, Knezevich A, et al. Monitoring tobacco-specific N-nitrosamines and nicotine in novel marlboro and camel smokeless tobacco products: findings from round I of the new product watch. *Nicotine Tob Res.* 2012;14(3):274–281.
24. Stepanov I, Gupta PC, Dhumal G, et al. High levels of tobacco-specific *N*-nitrosamines and nicotine in Chaini Khaini, a product marketed as snus. *Tob Control.* 2014. https://doi.org/10.1136/tobaccocontrol-2014-051744.

25. Hoffmann D, Hecht SS, Ornaf RM, Wynder EL, Tso TC. Nitrosonornicotine: presence in tobacco, formation and carcinogenicity. In: Walker EA, Bogovski P, Griciute L, eds. *Environmental N-Nitroso Compounds: Analysis and Formation*. 1st ed. Lyon, France: International Agency for Research on Cancer; 1976:307–320.
26. Hecht SS, Chen CB, Dong M, Ornaf RM, Hoffmann D, Tso TC. Studies on nonvolatile nitrosamines in tobacco. *Beitr Tabakforsch Int*. 1977;9:1–6.
27. Tobacco IARC. *Smoke and Involuntary Smoking. IARC Monographs on the Evaluation of Carcinogenic Risks to Humans*. Vol. 83. Lyon: FR: IARC; 2004.
28. Osterdahl BG, Jansson C, Paccou A. Decreased levels of tobacco-specific N-nitrosamines in moist snuff on the Swedish market. *J Agric Food Chem*. 2004;52(16): 5085–5088.
29. Idris AM, Nair J, Ohshima H, et al. Unusually high levels of carcinogenic tobacco-specific nitrosamines in Sudan snuff (toombak). *Carcinogenesis*. 1991;12:1115–1118.
30. Stepanov I, Hecht SS, Ramakrishnan S, Gupta PC. Tobacco-specific nitrosamines in smokeless tobacco products marketed in India. *Int J Cancer*. 2005;116:16–19.
31. Rutqvist LE, Curvall M, Hassler T, Ringberger T, Wahlberh I. Swedish snus and the GothiaTek standard. *Harm Reduct J*. 2011;8:11.
32. Lawler TS, Stanfill SB, Zhang L, Ashley DL, Watson CH. Chemical characterization of domestic oral tobacco products: total nicotine, pH, unprotonated nicotine and tobacco-specific *N*-nitrosamines. *Food Chem Toxicol*. 2013;57:380–386.
33. Hoffmann D, Djordjevic MV, Fan J, Zang E, Glynn T, Connolly GN. Five leading U.S. commercial brands of moist snuff in 1994 - assessment of carcinogenic *N*-nitrosamines. *J Natl Cancer Inst (Bethesda)*. 1995;87:1862–1869.
34. IARC. Some non-heterocyclic polycyclic aromatic hydrocarbons and some related exposures. *IARC Monographs on the Evaluation of Carcinogenic Risks to Humans*. 2010; 92. Lyon, France.
35. Uno S, Sakurai K, Nebert DW, Makishima M. Protective role of cytochrome P450 1A1 (CYP1A1) against benzo[a]pyrene-induced toxicity in mouse aorta. *Toxicology*. 2014;316:34–42.
36. IARC. Re-evaluation of some organic chemicals, hydrazine and hydrogen peroxide (part two). *IARC Monographs on the Evaluation of Carcinogenic Risks to Humans*. 1999: 318–335. Lyon, FR: IARC.
37. IARC. *A Review of Human Carcinogens: Chemical Agents and Related Occupations*. In: *IARC Monographs on the Evaluation of Carcinogenic Risks to Humans*. Vol. 100F. 2012. Lyon, France.
38. Brunnemann KD, Hoffmann D. *Chemical Composition of Smokeless Tobacco Products*. National Cancer Institute; 1992.
39. IARC. *Re-evaluation of Some Organic Chemicals, Hydrazine and Hydrogen Peroxide*. In: *IARC Monographs on the Evaluation of the Carcinogenic Risk of Chemicals to Humans*. Vol. 71. Lyon: FR: IARC; 1999.
40. IARC. *Dry cleaning, Some Chlorinated Solvents and other Industrial Chemicals*. In: *IARC Monographs on the Evaluation of Carcinogenic Risks to Humans*. Vol. 63. Lyon: France: IARC; 1995:373–391.
41. Adamu CA, Bell RE, Mulchi CL, Chanev RL. Residual metal levels in soils and leaf accumulations in tobacco a decade following farmland application of municipal sludge. *Environ Pollut*. 1989;56(2):113–126.
42. Mulchi CL, Adamu CA, Bell PF, Chaney RL. Residual heavy metal concentrations in sludge amended coastal plain soils - II. Predicting metal concentrations in tobacco from soil test information. *Commun Soil Sci Plant Anal*. 1992;23(9–10):1053–1069.
43. IARC. A Review of Human Carcinogens: Arsenic, Metals, Fibres, and Dusts. IARC Monographs on the Evaluation of Carcinogenic Risks to Humans, Vol. 100C. Lyon, France 2012.

44. Jarvis MJ, Fidler J, Mindell J, Feyerabend C, West R. Assessing smoking status in children, adolescents and adults: cotinine cut-points revisited. *Addiction*. 2008;103(9): 1553−1561.
45. Benowitz NL, Bernert JT, Caraballo RS, Holiday DB, Wang J. Optimal serum cotinine levels for distinguishing cigarette smokers and nonsmokers within different racial/ ethnic groups in the United States between 1999 and 2004. *Am J Epidemiol*. 2009; 169(2):236−248.
46. Hecht SS, Yuan JM, Hatsukami D. Applying tobacco carcinogen and toxicant biomarkers in product regulation and cancer prevention. *Chem Res Toxicol*. 2010;23: 1001−1008.
47. Yuan JM, Butler LM, Stepanov I, Hecht SS. Urinary tobacco smoke-constituent biomarkers for assessing risk of lung cancer. *Cancer Res*. 2014;74(2):401−411.
48. Park SL, Carmella SG, Ming X, et al. Variation in levels of the lung carcinogen NNAL and its glucuronides in the urine of cigarette smokers from five ethnic groups with differing risks for lung cancer. *Cancer Epidemiol Biomark Prev*. 2015;24(3):561−569.
49. Hecht SS, Carmella SG, Kotandeniya D, et al. Evaluation of toxicant and carcinogen metabolites in the urine of e-cigarette users versus cigarette smokers. *Nicotine Tob Res*. 2015;17(6):704−709.
50. Park SL, Carmella SG, Chen M, et al. Mercpaturic acids derived from the toxicants acrolein and crotonaldehyde in the urine of cigarettes smokers from five ethnic groups with differing risks for lung cancer. *PLoS One*. 2015;10:e0124841.
51. Chang CM, Rostron BL, Chang JT, et al. Biomarkers of exposure among U.S. Adult cigar smokers: population assessment of tobacco and health (PATH) study wave 1 (2013−2014). *Cancer Epidemiol Biomark Prev*. 2019;28(5):943−953.
52. Stepanov I, Jensen J, Hatsukami D, Hecht SS. Mass spectrometric quantitation of nicotine, cotinine, and 4-(methylnitrosamino)-1-(3-pyridyl)-1-butanol, in human toenails. *Cancer Epidemiol Biomark Prev*. 2006;15:2378−2383.
53. Murphy SE, Park S-SL, Thompson EF, et al. Nicotine *N*-glucurionidation relative to *N*-oxidation and *C*-oxidation and UGT2B10 genotype in five ethnic/racial groups. *Carcinogenesis*. 2014;35:2526−2533.
54. Hukkanen J, Jacob III P, Benowitz NL. Metabolism and disposition kinetics of nicotine. *Pharmacol Rev*. 2005;57(1):79−115.
55. Park SL, Carmella SG, Ming X, Stram DO, Le Marchand L, Hecht SS. Variation in levels of the lung carcinogen NNAL and its glucuronides in the urine of cigarette smokers from five ethnic groups with differing risks for lung cancer. *Cancer Epidemiol Biomark Prev*. 2015;24:561−569.
56. Patel YM, Park SL, Carmella SG, et al. Metabolites of the polycyclic aromatic hydrocarbon phenanthrene in the urine of cigarette smokers from five ethnic groups with differing risks for lung cancer. *PLoS One*. 2016;11(6):e0156203.
57. Haiman CA, Patel YM, Stram DO, et al. Benzene uptake and glutathione S-transferase T1 status as determinants of *S*-phenylmercapturic acid in cigarette smokers in the Multiethnic Cohort. *PLoS One*. 2016;11(3):e0150641.
58. Roethig HJ, Munjal S, Feng S, et al. Population estimates for biomarkers of exposure to cigarette smoke in adult U.S. cigarette smokers. *Nicotine Tob Res*. 2009;11(10): 1216−1225.
59. Shepperd CJ, Eldridge AC, Mariner DC, McEwan M, Errington G, Dixon M. A study to estimate and correlate cigarette smoke exposure in smokers in Germany as determined by filter analysis and biomarkers of exposure. *Regul Toxicol Pharmacol*. 2009; 55(1):97−109.
60. Alwis KU, deCastro BR, Morrow JC, Blount BC. Acrolein exposure in U.S. tobacco smokers and non-tobacco users: nhanes 2005−2006. *Environ Health Perspect*. 2015; 123(12):1302−1308.

61. Wei B, Blount BC, Xia B, Wang L. Assessing exposure to tobacco-specific carcinogen NNK using its urinary metabolite NNAL measured in US population: 2011–2012. *J Expo Sci Environ Epidemiol.* 2015;26(3):249–256.
62. Hecht SS, Stepanov I, Carmella SG. Exposure and metabolic activation biomarkers of carcinogenic tobacco-specific nitrosamines. *Acc Chem Res.* 2016;49(1):106–114.
63. Messina ES, Tyndale RF, Sellers EM. A major role for CYP2A6 in nicotine c-oxidation by human liver microsomes. *J Pharmacol Exp Ther.* 1997;282:1608–1614.
64. Benowitz NL, Pomerleau OF, Pomerleau CS, Jacob III P. Nicotine metabolite ratio as a predictor of cigarette consumption. *Nicotine Tob Res.* 2003;5(5):621–624.
65. Stellman SD, Chen Y, Muscat JE, et al. Lung cancer risk in white and black Americans. *Ann Epidemiol.* 2003;13(4):294–302.
66. Berg JZ, Mason J, Boettcher AJ, Hatsukami DK, Murphy SE. Nicotine metabolism in African Americans and European Americans: variation in glucuronidation by ethnicity and UGT2B10 haplotype. *J Pharmacol Exp Ther.* 2010;332(1):202–209.
67. Benowitz NL, Perez-Stable EJ, Herrera B, Jacob III P. Slower metabolism and reduced intake of nicotine from cigarette smoking in Chinese-Americans. *J Natl Cancer Inst (Bethesda).* 2002;94(2):108–115.
68. Hecht SS, Carmella SG, Foiles PG, Murphy SE. Biomarkers for human uptake and metabolic activation of tobacco-specific nitrosamines. *Cancer Res.* 1994;54:1912s–1917s.
69. Stepanov I, Hecht SS. Tobacco-specific nitrosamines and their *N*-glucuronides in the urine of smokers and smokeless tobacco users. *Cancer Epidemiol Biomark Prev.* 2005;14: 885–891.
70. Appleton S, Olegario RM, Lipowicz PJ. TSNA exposure from cigarette smoking: 18 years of urinary NNAL excretion data. *Regul Toxicol Pharmacol.* 2014;68(2):269–274.
71. Vogel RI, Carmella SG, Stepanov I, Hatsukami DK, Hecht SS. The ratio of a urinary tobacco-specific lung carcinogen metabolite to cotinine is significantly higher in passive than in active smokers. *Biomarkers.* 2011;16:491–497.
72. Stepanov I, Sebero E, Wang R, Gao YT, Hecht SS, Yuan JM. Tobacco-specific *N*-nitrosamine exposures and cancer risk in the Shanghai cohort study: remarkable coherence with rat tumor sites. *Int J Cancer.* 2013;134:2278–2283.
73. Hecht SS, Carmella SG, Chen M, et al. Quantitation of urinary metabolites of a tobacco-specific lung carcinogen after smoking cessation. *Cancer Res.* 1999;59: 590–596.
74. Goniewicz ML, Havel CM, Peng MW, et al. Elimination kinetics of the tobacco-specific biomarker and lung carcinogen 4-(methylnitrosamino)-1-(3-pyridyl)-1-butanol. *Cancer Epidemiol Biomark Prev.* 2009;18(12):3421–3425.
75. Stepanov I, Carmella SG, Briggs A, et al. Presence of the carcinogen *N'*-nitrosonornicotine in the urine of some users of oral nicotine replacement therapy products. *Cancer Res.* 2009;69(21):8236–8240.
76. Hochalter JB, Zhong Y, Han S, Carmella SG, Hecht SS. Quantitation of a minor enantiomer of phenanthrene tetraol in human urine: correlations with levels of overall phenanthrene tetraol, benzo[*a*]pyrene tetraol, and 1-hydroxypyrene. *Chem Res Toxicol.* 2011;24:262–268.
77. Hecht SS, Carmella SG, Villalta PW, Hochalter JB. Analysis of phenanthrene and benzo[*a*]pyrene tetraol enantiomers in human urine: relevance to the bay region diol epoxide hypothesis of benzo[*a*]pyrene carcinogenesis and to biomarker studies. *Chem Res Toxicol.* 2010;23(5):900–908.
78. Hecht SS, Hochalter JB, Carmella SG, et al. Longitudinal study of [D_{10}]phenanthrene metabolism by the diol epoxide pathway in smokers. *Biomarkers.* 2013;18(2):144–150.
79. Wang J, Zhong Y, Carmella SG, et al. Phenanthrene metabolism in smokers: use of a two-step diagnostic plot approach to identify subjects with extensive metabolic activation. *J Pharmacol Exp Ther.* 2012;342(3):750–760.

80. Hecht SS, Chen M, Yagi H, Jerina DM, Carmella SG. *r*-1,*t*-2,3,*c*-4-Tetrahydroxy-1,2,3,4-tetrahydrophenanthrene in human urine: a potential biomarker for assessing polycyclic aromatic hydrocarbon metabolic activation. *Cancer Epidemiol Biomark Prev*. 2003;12:1501–1508.
81. Carmella SG, Chen M, Yagi H, Jerina DM, Hecht SS. Analysis of phenanthrols in human urine by gas chromatography-mass spectrometry: potential use in carcinogen metabolite phenotyping. *Cancer Epidemiol Biomark Prev*. 2004;13:2167–2174.
82. Fant RV, Henningfield JE, Nelson RA, Pickworth WB. Pharmacokinetics and pharmakodynamics of moist snuff in humans. *Tob Control*. 1999;8:387–392.
83. Hecht SS, Carmella SG, Murphy SE, et al. Similar exposure to a tobacco-specific carcinogen in smokeless tobacco users and cigarette smokers. *Cancer Epidemiol Biomark Prev*. 2007;16(8):1567–1572.
84. Benowitz NL, Renner CC, Lanier AP, et al. Exposure to nicotine and carcinogens among Southwestern Alaskan Native cigarette smokers and smokeless tobacco users. *Cancer Epidemiol Biomark Prev*. 2012;21(6):934–942.
85. Lemmonds CA, Hecht SS, Jensen JA, et al. Smokeless tobacco topography and toxin exposure. *Nicotine Tob Res*. 2005;7:469–474.
86. Hecht SS, Carmella SG, Ye M, et al. Quantitation of metabolites of 4-(methylnitrosamino)-1-(3-pyridyl)-1-butanone after cessation of smokeless tobacco use. *Cancer Res*. 2002;62:129–134.
87. Hatsukami DK, Anton D, Callies A, Keenan R. Situational factors and patterns associated with smokeless tobacco use. *J Behav Med*. 1991;14(4):383–396.
88. Hatsukami DK, Gust SW, Keenan RM. Physiologic and subjective changes from smokeless tobacco withdrawal. *Clin Pharmacol Ther*. 1987;41(1):103–107.
89. Hatsukami DK, Keenan RM, Anton DJ. Topographical features of smokeless tobacco use. *Psychopharmacology (Berlin)*. 1988;96(3):428–429.
90. Severson HH, Eakin EG, Lichtenstein E, Stevens VJ. The inside scoop on the stuff called snuff: an interview study of 94 adult male smokeless tobacco users. *J Subst Abus*. 1990;2(1):77–85.
91. Hatsukami DK, Feuer RM, Ebbert JO, Stepanov I, Hecht SS. Changing smokeless tobacco products: new tobacco delivery systems. *Am J Prev Med*. 2007;33:S368–S378.
92. Hatsukami DK, Stepanov I, Severson H, et al. Evidence supporting product standards for carcinogens in smokeless tobacco. *Cancer Prev Res*. 2014 (in press).
93. Hatsukami DK, Lemmonds C, Zhang Y, et al. Evaluation of carcinogen exposure in people who used "reduced exposure" tobacco products. *J Natl Cancer Inst (Bethesda)*. 2004;96:844–852.
94. Yuan JM, Knezevich AD, Wang R, Gao YT, Hecht SS, Stepanov I. Urinary levels of the tobacco-specific carcinogen N'-nitrosonornicotine and its glucuronide are strongly associated with esophageal cancer risk in smokers. *Carcinogenesis*. 2011; 32(9):1366–1371.
95. Yuan JM, Gao YT, Murphy SE, et al. Urinary levels of cigarette smoke constituent metabolites are prospectively associated with lung cancer development in smokers. *Cancer Res*. 2011;71(21):6749–6757.
96. Church TR, Anderson KE, Caporaso NE, et al. A prospectively measured serum biomarker for a tobacco-specific carcinogen and lung cancer in smokers. *Cancer Epidemiol Biomark Prev*. 2009;19:260–266.
97. Yuan JM, Koh WP, Murphy SE, et al. Urinary levels of tobacco-specific nitrosamine metabolites in relation to lung cancer development in two prospective cohorts of cigarette smokers. *Cancer Res*. 2009;69(7):2990–2995.
98. Kresty LA, Carmella SG, Borukhova A, et al. Metabolites of a tobacco-specific nitrosamine, 4-(methylnitrosamino)-1-(3-pyridyl)-1-butanone (NNK), in the urine of

smokeless tobacco users: relationship of urinary biomarkers and oral leukoplakia. *Cancer Epidemiol Biomark Prev.* 1996;5:521–525.

99. Murphy SE, Villalta P, Ho SW, von Weymarn LB. Analysis of [3',3'-d(2)]-nicotine and [3',3'-d(2)]-cotinine by capillary liquid chromatography-electrospray tandem mass spectrometry. *J Chromatogr B Analyt Technol Biomed Life Sci.* 2007;857:1–8.
100. Yue B, Kushnir MM, Urry FM, Rockwood AL. *Quantitation of Nicotine, Its Metabolites, and other Related Alkaloids in Urine, Serum, and Plasma Using LC-MS-MS.* In: *Clinical Applications of Mass Spectrometry*. Humana Press; 2010:389–398.
101. Hoffmann D, Adams JD. Carcinogenic tobacco-specific N-nitrosamines in snuff and saliva of snuff-dippers. *Cancer Res.* 1981;41:4305–4308.
102. Harris AC, Tally L, Schmidt CE, et al. Animal models to assess the abuse liability of tobacco products: effects of smokeless tobacco extracts on intracranial self-stimulation. *Drug Alcohol Depend.* 2015;147:60–67.
103. Stepanov I, Knezevich A, Zhang L, Watson CH, Hatsukami DK, Hecht SS. Carcinogenic tobacco-specific N-nitrosamines in US cigarettes: three decades of remarkable neglect by the tobacco industry. *Tob Control.* 2012;21:44–48.
104. Wu W, Ashley DL, Watson CH. Simultaneous determination of five tobacco-specific nitrosamines in mainstream cigarette smoke by isotope dilution liquid chromatography/electrospray ionization tandem mass spectrometry. *Anal Chem.* 2003;75: 4827–4832.
105. Ding YS, Trommel JS, Yan XJ, Ashley D, Watson CH. Determination of 14 polycyclic aromatic hydrocarbons in mainstream smoke from domestic cigarettes. *Environ Sci Technol.* 2005;39(2):471–478.
106. Ding YS, Ashley DL, Watson CH. Determination of 10 carcinogenic polycyclic aromatic hydrocarbons in mainstream cigarette smoke. *J Agric Food Chem.* 2007;55: 5966–5973.
107. Borgerding MF, Bodnar JA, Wingate DE. *The 1999 Massachusetts Benchmark Study - Final Report. A Research Study Conducted after Consultation with the Massachusetts Department of Public Health*; 2000. http://www_brownandwilliamson_com/APPS/PDF/Final_Report_1999_Mass_Benchmark_Study_pdf.
108. Counts ME, Hsu FS, Laffoon SW, Dwyer RW, Cox RH. Mainstream smoke constituent yields and predicting relationships from a worldwide market sample of cigarette brands: ISO smoking conditions. *Regul Toxicol Pharmacol.* 2004;39:111–134.
109. Houlgate PR, Dhingra KS, Nash SJ, Evans WH. Determination of formaldehyde and acetaldehyde in mainstream cigarette smoke by high-performance liquid chromatography. *Analyst (London).* 1989;114:355–360.
110. Ding YS, Yan X, Wong J, Chan M, Watson CH. In situ derivatization and quantification of seven carbonyls in cigarette mainstream smoke. *Chem Res Toxicol.* 2016; 29(1):125–131.
111. Fresquez MR, Pappas RS, Watson CH. Establishment of toxic metal reference range in tobacco from US cigarettes. *J Anal Toxicol.* 2013;37(5):298–304.
112. Stepanov I, Hecht SS. Tobacco-specific nitrosamines and their pyridine-N-glucuronides in the urine of smokers and smokeless tobacco users. *Cancer Epidemiol Biomarkers Prev.* 2005;14(4):885–891.
113. Knezevich A, Muzic J, Hatsukami DK, Hecht SS, Stepanov I. Nornicotine nitrosation in saliva and its relation to endogenous synthesis of N'-nitrosonornicotine in humans. *Nicotine Tob Res.* 2013;15(2):591–595.
114. Bustamante G, Yakovlev G, Yershova K, et al. *Endogenous Formation of the Oral and Esophageal Carcinogen N'-nitrosonornicotine in e-cigarette Users.* Cancer Research; 2017 (submitted manuscript).
115. Carmella SG, Ming X, Olvera N, Brookmeyer C, Yoder A, Hecht SS. High throughput liquid and gas chromatography-tandem mass spectrometry assays for

tobacco-specific nitrosamine and polycyclic aromatic hydrocarbon metabolites associated with lung cancer in smokers. *Chem Res Toxicol.* 2013;26(8):1209–1217.

116. Kotandeniya D, Carmella SG, Ming X, Murphy SE, Hecht SS. Combined analysis of the tobacco metabolites cotinine and 4-(Methylnitrosamino)-1-(3-pyridyl)-1-butanol in human urine. *Anal Chem.* 2015;87(3):1514–1517.
117. Hecht SS, Carmella SG, Le K, et al. Effects of reduced cigarette smoking on levels of 1-hydroxypyrene in urine. *Cancer Epidemiol Biomark Prev.* 2004;13:834–842.
118. Khariwala SS, Carmella SG, Stepanov I, et al. Elevated levels of 1-hydroxypyrene and N'-nitrosonornicotine in smokers with head and neck cancer: a matched control study. *Head Neck.* 2013;35(8):1096–1100.
119. Carmella SG, Chen M, Han S, et al. Effects of smoking cessation on eight urinary tobacco carcinogen and toxicant biomarkers. *Chem Res Toxicol.* 2009;22(4):734–741.
120. Caldwell KL, Hartel J, Jarrett J, Jones RL. Inductively coupled plasma mass spectrometry to measure multiple toxic elements in urine in NHANES 1999-2000. *Atom Spectrosc.* 2005;26(1):1–7.
121. Kim K, Steuerwald AJ, Parsons PJ, Fujimoto VY, Browne RW, Bloom MS. Biomonitoring for exposure to multiple trace elements via analysis of urine from participants in the Study of Metals and Assisted Reproductive Technologies (SMART). *J Environ Monit.* 2011;13:2413–2419.
122. Minnich MG, Miller DC, Parsons PJ. Determination of As, Cd, Pb, and Hg in urine using inductively coupled plasma mass spectrometry with the direct injection high efficiency nebulizer. *Spectrochim Acta Part B At Spectrosc.* 2008;63:389–395.
123. WHO. *WHO TobReg: Report on the Scientific Basis of Tobacco Product Regulation: 5th Report of a WHO Study Group*. In: *WHO Technical Report Series*. Vol. 989. Geneva: World Health Organization, Study Group on Tobacco product Regulation; 2015. https://www.who.int/tobacco/publications/prod_regulation/trs989/en/.

CHAPTER SEVEN

Metabolism and DNA adduct formation of carcinogenic tobacco-specific nitrosamines found in smokeless tobacco products

Stephen S. Hecht
University of Minnesota, Minneapolis, MN, United States

Introduction

All commercially available smokeless tobacco products are contaminated with carcinogenic, tobacco-specific nitrosamines in amounts that exceed by 100–1000 times the levels of potentially dangerous compounds in human food or beverages.[1–8] Some other carcinogens, including traces of polycyclic aromatic hydrocarbons and volatile aldehydes, certain lactones, and metals, are also present in smokeless tobacco. However, the tobacco-specific nitrosamines and nitrosamino acids are consistently the most abundant. Magee and Barnes first established the powerful carcinogenicity of *N*-nitrosodimethylamine (NDMA), the simplest dialkylnitrosamine, in 1956 (Fig. 7.1).[9] Years later, extensive dose–response studies of NDMA and *N*-nitrosodiethylamine on more than 4000 rats established a linear relationship between dose and carcinogenicity down to levels of only 0.1 ppm of the carcinogens in the drinking water, without any indication of a threshold.[10] (Fig. 7.1)

Multiple studies carried out in the second half of the 20th century also demonstrated the generality of nitrosamine carcinogenicity, with over 200

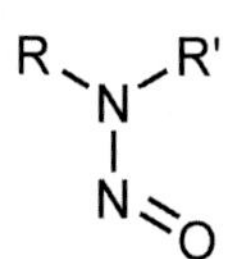

Figure 7.1 General structure of a dialkylnitrosamine.

Smokeless Tobacco Products
ISBN: 978-0-12-818158-4
https://doi.org/10.1016/B978-0-12-818158-4.00007-8

nitrosamines of various substitution patterns showing clear carcinogenicity in at least 30 species ranging from mollusks to primates, thus establishing nitrosamines as a major class of highly carcinogenic compounds.[11–14] Important among these, particularly in view of human exposure levels, are the tobacco-specific nitrosamines, genotoxic carcinogens which are virtually unique to tobacco products and are found in substantial amounts in all commercially available smokeless tobacco products. The first identification of *N*′-nitrosonornicotine (NNN) in smokeless tobacco was reported in 1974; this was also the first report of any carcinogen in unburned tobacco.[15] 4-(Methylnitrosamino)-1-(3-pyridyl)-1-butanone (NNK) was first detected in smokeless tobacco in 1978.[16] Ultimately, seven tobacco-specific nitrosamines—NNN, NNK, 4-(methylnitrosamino)-1-(3-pyridyl)-1-butanol (NNAL), 4-(methylnitrosamino)-4-(3-pyridyl)-1-butanol (*iso*-NNAL), 4-(methylnitrosamino)-4-(3-pyridyl)butyric acid (*iso*-NNAC), *N*′-nitrosoanatabine (NAT), and *N*′-nitrosoanabasine (NAB)—were identified in smokeless tobacco (Fig. 7.2).[1] NNN and NNK have received by far the most attention because of their potent carcinogenic activities and occurrence in smokeless tobacco products. NNN and NNK, which always occur together in smokeless tobacco, are considered "carcinogenic to humans" by the International Agency for Research on Cancer.[1] The major NNK metabolite NNAL, a strong carcinogen with activity similar to that of NNK, also has been extensively studied, although its levels in tobacco are lower than those of NNN and NNK.[3,6] NAT and NAB also have been routinely measured in smokeless tobacco products. NAT levels are typically similar to those of NNN, but carcinogenicity assays of NAT are negative, while NAB is only weakly carcinogenic.[3,6] There is no evidence for carcinogenicity of *iso*-NNAL and *iso*-NNAC.[17] These carcinogenicity data are based on comparative bioassays in F-344 rats, A/J mice, and other laboratory animals.[17] This chapter will focus

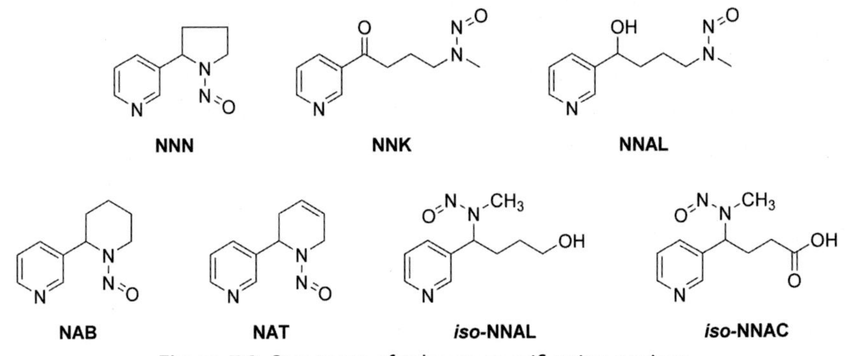

Figure 7.2 Structures of tobacco-specific nitrosamines.

on NNN, NNK, and NNAL, summarizing carcinogenicity data, metabolism pathways and their relationship to carcinogenicity, metabolite biomarkers for human exposure, and DNA and protein adduct formation. In Chapter 7, Stepanov and Hatsukami summarize levels of tobacco-specific nitrosamines in smokeless tobacco products. Collectively, the available data based on more than 40 years of intensive research by academia, government agencies, and the tobacco industry clearly demonstrate that tobacco-specific nitrosamines are the most prevalent strong carcinogens in smokeless tobacco products. Consequently, the US Food and Drug Administration (FDA) has proposed regulation of NNN levels in these products, not to exceed 1 part per million (based on dry weight).

Carcinogenicity of NNN, NNK, and NNAL

A detailed review of carcinogenicity studies of NNN, NNK, and NNAL through 1997 has been published.[17] Some highlights of that extensive data set are presented here, along with relevant new data.

The oral mucosa, esophagus, and nasal mucosa are the main target tissues of NNN in the rat, and depend significantly on the route of administration. When NNN is given in the drinking water, tumors are observed mainly in the tissues of contact—oral mucosa and esophagus.[17] The lowest dose to have produced esophageal tumors in rats was 5 ppm in the drinking water.[18] When NNN is administered by subcutaneous (SC) injection, nasal tumors are predominant. As discussed below, these tissues are sites of efficient metabolism and consequent DNA binding of NNN in the rat. Tracheal and nasal tumors are mainly induced by administration of NNN to Syrian golden hamsters, independent of the route of administration. NNN consistently induces tumors of the lung in various mouse strains, but the effect is relatively weak. The mink is the only nonrodent model in which NNN has been tested; it induced predominantly nasal mucosa tumors, primarily esthesioneuroepitheliomas with invasion of the brain.[17]

The lung is the main target of NNK administered to rats. Lung tumors are induced in F-344 rats independently of the route of administration of NNK—drinking water, SC injection, gavage, oral swabbing, or intravesicular administration.[17] Dose—response studies demonstrated that a total dose of only 1.8 mg/kg, administered by SC injection, was sufficient to induce a significant incidence of lung tumors. NNAL has similar carcinogenicity as NNK in the rat. Nasal tumors are the second most common neoplasm in rats treated with NNK by SC injection. Liver tumors and pancreatic tumors have also been observed. Sensitive and resistant strains of mice all develop lung tumors when treated with NNK, and the A/J mouse, a sensitive strain, is a commonly used model for lung tumor induction using NNK. Lung, trachea, and nasal mucosa are the main target tissues of NNK in the Syrian golden hamster. Overall, the organoselectivity of NNK for the lung in all models tested is a remarkable feature of this nitrosamine.[17]

Two relatively recent studies provide important new data with respect to the carcinogenicity of NNN, NNK, and NNAL. One study examined the carcinogenicity of the enantiomers of NNN, administered in the drinking water to F-344 rats.[19] (*S*)-NNN is the major NNN enantiomer in smokeless tobacco products, comprising 57%–66% of total NNN in these products.[5] Groups of rats were treated with either (*S*)-NNN or (*R*)-NNN at a dose of 14 ppm in the drinking water, or with racemic NNN at a dose of 28 ppm. The experiment was terminated after 17 months for (*S*)-NNN and racemic NNN, and after 20 months for (*R*)-NNN and untreated rats. At necropsy, all rats treated with (*S*)-NNN had oral cavity tumors. A total of 89 oral cavity tumors were observed in the 20 rats necropsied, including tumors of the tongue, buccal and gingival oral mucosa, soft palate, epiglottis, and pharynx (Fig. 7.3). Some of the tumors were large.

The rats treated with (*S*)-NNN also had 122 esophageal tumors. (*R*)-NNN was by comparison relatively weak, inducing only six oral cavity tumors and three esophageal tumors in the 24 rats necropsied. Racemic NNN was also highly carcinogenic to the rat oral cavity and esophagus. All rats treated with racemic NNN had these tumors. A total of 96 oral cavity tumors and 153 esophageal tumors were observed in the 12 rats necropsied. Thus the carcinogenicity of racemic NNN in the oral cavity and esophagus was significantly greater than that of (*S*)-NNN or than the additive effects of (*R*)-NNN and (*S*)-NNN, indicating that (*R*)-NNN was a cocarcinogen, synergistically enhancing the carcinogenicity of (*S*)-NNN. This study thus identified (*S*)-NNN as the only powerful oral mucosa carcinogen in smokeless tobacco products, strongly implicating its role in causing oral cavity cancer in smokeless tobacco users and supporting the FDA's proposal for regulation of NNN in these products.

The second study compared the carcinogenic activities of NNK (5 ppm), (*S*)-NNAL (5 ppm), (*R*)-NNAL (5 ppm), and racemic NNAL (10 ppm) administered in the drinking water to F-344 rats.[20] Both NNAL enantiomers are formed to varying extents in human tissues exposed to NNK. All of the compounds induced a high incidence of lung tumors, both adenomas and carcinomas. NNK and racemic NNAL were the most potent;

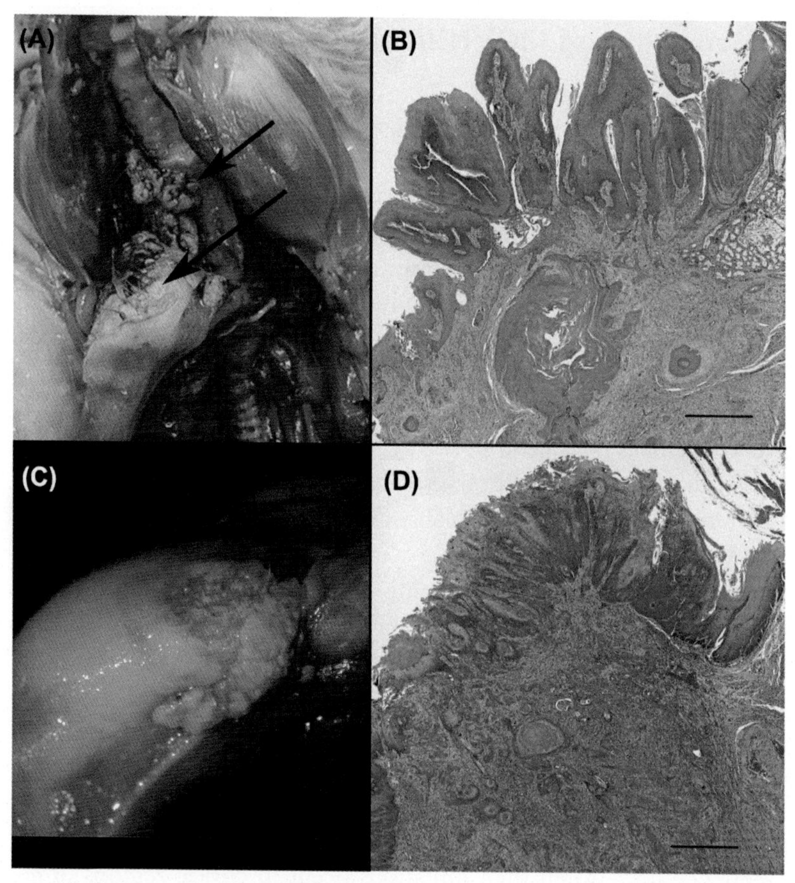

Figure 7.3 Squamous cell carcinoma in F-344 rats treated with (*S*)-NNN: (A) squamous cell carcinomas involving both the soft palate (short arrow) and tongue (long arrow); (B) corresponding histology showing invasion of soft palate (bar = 500 μm); (C) squamous cell carcinoma involving the tongue; (D) histology showing invasion of underlying skeletal muscle.

(*S*)-NNAL and (*R*)-NNAL had equivalent activities. Nearly all of the treated rats had both types of lung tumors; some of the tumors were very large. Pancreatic tumors were also observed and were determined to be due to metastasis from the lung tumors. Thus, this study confirmed the powerful pulmonary carcinogenicity of NNK and both NNAL enantiomers. The relevance of these results to smokeless tobacco use was recently demonstrated in the Agricultural Health Study, the results of which showed that exclusive long-term users of smokeless tobacco had an elevated risk of lung cancer, which is logically attributable to NNK and NNAL exposure.[21]

Metabolism of NNK, NNAL, and NNN in rodents and primates

Metabolism is the key to understanding the effects of carcinogenic nitrosamines such as NNK, NNAL, and NNN. All nitrosamines require metabolism to exert their carcinogenic effects. The nitrosamines themselves are unreactive, stable compounds. Cytochrome P450 enzymes present in human tissues catalyze the metabolism of nitrosamines, as illustrated in Fig. 7.4 for the simplest dialkylnitrosamine, NDMA.[11] The critical reaction is hydroxylation of the carbon atom adjacent (alpha-) to the *N*-nitroso-group, which results in the generation of an unstable α-hydroxynitrosamine (**1**), with a half-life typically measured in seconds. The α-hydroxynitrosamine spontaneously decomposes to a diazohydroxide intermediate **2**, then to a diazonium ion **3**, both of which are also highly unstable. These latter intermediates are electrophilic and readily react with nucleophilic sites in DNA, RNA, and protein, along with their predominant reaction with H_2O to produce alcohol-containing metabolites. The reactions with DNA are critical in carcinogenesis because they produce DNA addition products (adducts) that can cause mutations during DNA replication. The classic example of this scenario is the formation of O^6-methylguanine (O^6-MeG, **4**) from guanine during the metabolism of NDMA. O^6-MeG has miscoding properties and, if not repaired by the repair protein O^6-alkylguanine-DNA alkyltransferase, can be misread as T during replication, resulting in insertion of A, and a consequent permanent G → A mutation in the sequence.[22,23] When such mutations occur in critical regions of genes involved in growth control, such as oncogenes or tumor suppressor genes, the result can be initiation of the cancer process.

Applying this established mechanism to NNK and NNN results in the cascade of intermediates and products summarized in Fig. 7.5.[1,17,24] An additional feature that is specific to NNK is its metabolic reduction to NNAL which occurs readily in virtually all human and animal tissues studied

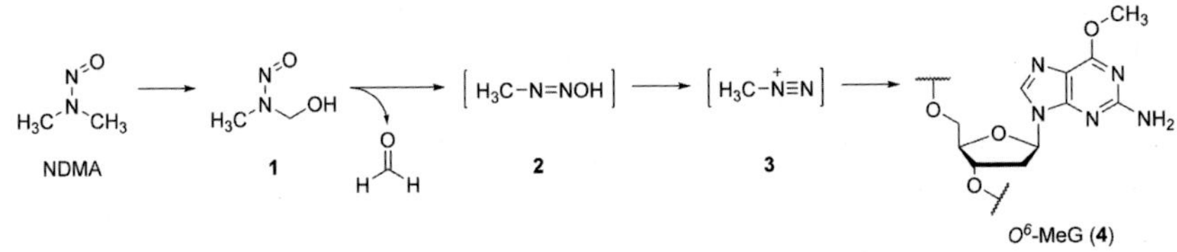

Figure 7.4 Metabolic activation of NDMA.

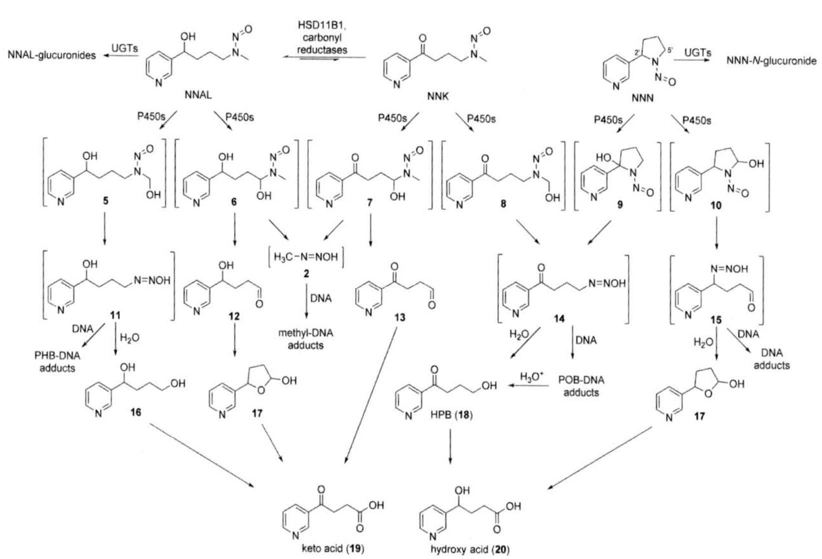

Figure 7.5 Metabolic pathways of NNK, NNAL, and NNN leading to DNA adduct formation and urinary metabolites.

to date. Cytochromes P450, particularly human enzymes P450 2A13 and P450 2A6, efficiently catalyze the α-hydroxylation of NNK and NNN, respectively, and P450 2A6 is also the primary catalyst of nicotine metabolism. High activity of P450 2A6 in smokers has been associated with higher lung cancer risk due to faster nicotine and NNK metabolism.[25] α-Hydroxylation of NNK produces the α-hydroxynitrosamines **7** and **8** while α-hydroxylation of NNN yields intermediates **9** and **10**. Similarly, α-hydroxylation of NNAL gives the unstable intermediates **5** and **6**.

Intermediates **6** and **7** formed from NNAL and NNK both spontaneously decompose to methyl diazohydroxide (**2**), the same reactive species produced during metabolism of NDMA. Intermediate **8** from NNK decomposes to diazohydroxide **14** while intermediate **5** from NNAL decomposes to diazohydroxide **11**. Diazohydroxide **14** is also produced by cytochrome P450 catalyzed 2′-hydroxylation of NNN, which initially yields compound **9**. 5′-Hydroxylation of NNN gives unstable intermediate **10**, which spontaneously ring opens to the aldehydic diazohydroxide **15**.

All diazohydroxides generated by these cytochrome P450 catalyzed reactions react with DNA to produce adducts: pyridylhydroxybutyl (PHB)-DNA adducts from **11**, methyl-DNA adducts from **2**, pyridyloxobutyl (POB)-DNA adducts from **14**, and NNN-5′-hydroxylation DNA adducts from **15**. The ultimate fate of the intermediary metabolites produced in these reactions—**12**, **13**, **16**, **17**, and **18**—is further metabolic oxidation to give keto acid **19** and hydroxy acid **20**, the major urinary metabolites of NNK, NNAL, and NNN, as determined from metabolism studies in rodents and primates.[1,17,24]

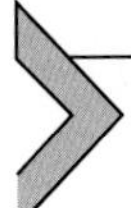

Tobacco-specific nitrosamines and their metabolites as biomarkers in smokeless tobacco users

Urine

Urinary total NNAL, the sum of free NNAL and its *O*- and *N*-glucuronides (NNAL-Gluc), metabolites of the carcinogen NNK, is presently the best biomarker of tobacco-specific carcinogen exposure in smokeless tobacco users. Advantages of this biomarker include its direct relationship to NNK exposure, its tobacco specificity, its relatively long lifetime, and the availability of practical highly sensitive, quantitative, robust, and artifact-free mass spectrometric methods for its measurement, applicable in large studies. NNN, NAB, and NAT as well as their *N*-glucuronides have also been detected in the urine of smokeless tobacco users, but the wide application of these biomarkers has not yet been realized because of challenges in their quantitation.[26] For NNN analysis in urine, [pyridine-D_4]nornicotine has been used as a tracer to monitor possible artifactual formation of NNN during analysis. In one study, this approach allowed determination of the enantiomeric composition of NNN in urine, which was 56 ± 3% (*S*)-NNN, the more carcinogenic isomer.[27]

Keto acid **19** and hydroxy acid **20** (Fig. 7.5) would also be potential urinary biomarkers of exposure and metabolic activation of NNK and NNN. However, these compounds are also minor metabolites of nicotine, which is present in smokeless tobacco in amounts more than 1000 times as great as NNK and NNN.[3] Therefore, their measurement in urine does not reflect NNK and NNN exposure because the amounts formed from these carcinogens are insignificant compared to the amounts produced in nicotine metabolism.

The first study to report levels of NNAL and its *O*-glucuronide in the urine of smokeless tobacco users quantified these substances in Sudanese snuff dippers who used "toombak," a local oral snuff product with exceptionally high levels of NNK (195–365 μg/g). The amounts found were approximately 1000-fold higher than observed in later studies of smokeless tobacco users in the United States.[28] Toombak is considered to play a major role as a cause of oral cancer in Sudan.[29]

Kresty et al. were the first to analyze levels of NNAL and NNAL-Gluc in the urine of smokeless tobacco users from the United States. They observed significant associations between the presence of leukoplakia and urinary levels of NNAL, NNAL-Gluc, and total NNAL.[30] In a study of urinary NNAL and NNAL-Gluc levels after cessation of smokeless tobacco use, the decay of these metabolites was faster than observed in smokers who stopped smoking, for reasons that are not entirely clear. In the same study, ratios of (*S*)-NNAL:(*R*)-NNAL and (*S*)-NNAL-Gluc:(*R*)-NNAL-Gluc were significantly higher 7 days after cessation than at baseline, suggesting selective retention in the body of (*S*)-NNAL.[31]

We compared levels of total NNAL in the urine of smokeless tobacco users and cigarette smokers.[32] Levels of total NNAL per milliliter of urine and per milligram creatinine were significantly higher in smokeless tobacco users than in cigarette smokers. These results demonstrated that smokeless tobacco is not a harmless substitute for cigarette smoking because there is still considerable carcinogen uptake, although the variety of carcinogens and exposure to the lung is far greater in cigarette smokers than in smokeless tobacco users. Rostron et al. obtained similar results in the NHANES study in which exclusive smokeless tobacco users had higher geometric mean concentrations of total NNAL than exclusive cigarette smokers.[33]

Further understanding of the NNAL biomarker in smokeless tobacco users was achieved by determining the percentage of the NNK dose that was excreted as total NNAL. In this study, the amount of NNK in a single dose of smokeless tobacco was measured before and after use, along with the amount in expectorated saliva, thus allowing calculation of NNK dose. Total NNAL in these subjects was measured over a 3-day period. The results showed that the percent conversion of NNK to total NNAL in smokeless tobacco users was 14%—17%.[34]

The relationship of duration and amount of smokeless tobacco use as well as constituent levels of NNK and NNN have been shown to be significant determinants of levels of total NNAL and total NNN in urine, supporting the need for product standards for carcinogens in smokeless tobacco products.[1,35]

Benowitz et al. compared levels of total NNAL and total NNN in the urine of Alaskan Native people who used different products. Levels of total NNAL were highest in users of smokeless tobacco, followed by those who used iqmik, a homemade Alaskan smokeless tobacco product prepared with dried tobacco leaves mixed with alkaline ash. Total NNN levels were similar in the two groups.[36]

Saliva

In international studies, NNK, NNN, NAB, and NAT have been detected in the saliva of users of various types of smokeless tobacco, in amounts that reflect their concentrations in tobacco. The highest levels have been reported in users of toombak and certain products from India.[1]

DNA and protein adduct formation from tobacco-specific nitrosamines

While metabolites in urine and other body fluids as discussed in the previous section provide an indication of carcinogen dose, the formation of DNA adducts, which are central in the carcinogenic process, will not necessarily correlate with dose in smokeless tobacco users because of individual differences in carcinogen metabolism (e.g., metabolic activation vs. detoxification) or differences in DNA repair capacity. Thus, the study of DNA adducts potentially can provide significant insights on individual risk for cancer in smokeless tobacco users.

DNA adducts are critical in carcinogenesis by tobacco-specific nitrosamines. Convincing data from studies in mice and rats demonstrate their presence and persistence in target tissues and support a role for their centrality in carcinogenesis by NNK and NNN.[20,22,37–40] DNA adducts are formed by the α-hydroxylation metabolism pathways shown in Fig. 7.5 as summarized in Table 7.1. While DNA adduct formation in laboratory animals treated with tobacco-specific nitrosamines has been extensively characterized, PHB-DNA adducts and POB-DNA adducts, which are specifically related to NNK and NNN, have yet to be identified in human subjects.

However, as shown in Fig. 7.5, acid hydrolysis of DNA isolated from tissues of rats treated with NNK or NNN releases 4-hydroxy-1-(3-pyridyl)-1-butanone (HPB, **18**).[17,41] Thus, HPB released from DNA can potentially serve as a biomarker for DNA adduct formation by tobacco-specific nitrosamines. This biomarker has been quantified in human DNA samples from oral cells and lung tissue, but studies of HPB-releasing DNA adducts in smokeless tobacco users have not been reported.[42–44] Considering that HPB-releasing hemoglobin adducts, discussed below, were elevated in snuff dippers, it would be important to carry out a similar study with oral cell DNA.[45]

Hemoglobin adducts have been proposed as internal dosimeters to assess metabolic activation of carcinogens on an individual basis.[46] Advantages of hemoglobin adducts include the ease of isolation of hemoglobin from blood

Table 7.1 Summary of DNA adduct formation from NNK, NNAL, and NNN as observed in laboratory animals treated with the carcinogens.

Parent compound	Initially formed metabolites in Fig. 7.4 or 7.5	Diazohydroxide intermediate in Fig. 7.5	Representative DNA adducts identified in vivo	Key references
NNK, NNAL	**6,7**	**2**	7-MeG, O^6-MedG, O^4-MeT, Me-DNA-phosphate	17
NNAL	**5**	**11**	7-PHB-G, O^6-PHB-dG, O^2-PHB-T, N^6-PHB-dA, PHB-DNA-phosphate	20,40,51,52
NNK, NNN	**8,9**	**14**	7-POB-G, O^6-POB-dG, O^2-POB-C, O^2-POB-T, POB-DNA-phosphate	20,37,51,53,54
NNN	**10**	**15**	Py,Py-dI	55

7-MeG, 7-methylguanine; O^6-MedG, O^6-methyldeoxyguanosine; O^4-MeT, O^4-methylthymidine; 7-PHB-G, 7-pyridylhydroxybutylguanine; N^6-PHB-dA; N^6-pyridylhydroxybutyldeoxyadenosine; 7-POB-G, 7-pyridyloxobutylguanine; O^2-POB-C, O^2-pyridyloxobutylcytosine; POB-DNA-phosphate, pyridyloxobutyl-DNA-phosphate adducts; Py,Py-dI, 2-(2-(3-pyridyl)-*N*-pyrrolidinyl)-2′-deoxyinosine.

and the relatively long lifetime of the erythrocyte (120 days in humans) potentially allowing estimation of chronic exposure plus metabolic activation of carcinogens. Treatment of rats with NNK or NNN results in the formation of hemoglobin adducts which can release HPB.[47] The released HPB can be quantified using mass spectrometric methods.[45] Studies have shown that intermediate **14** (Fig. 7.5) binds to aspartate or glutamate of globin, and that these ester adducts release HPB upon mild base treatment.[48] A correlation between HPB-releasing hemoglobin adducts and DNA adducts has been established in F-344 rats treated with a range of NNK doses.[41] Further studies in F-344 rats demonstrated that HPB-releasing hemoglobin adducts decrease significantly in rats treated with NNK plus 2-phenethyl isothiocyanate, an inhibitor of NNK-induced lung carcinogenesis, compared to rats treated only with NNK.[49] Validated methods for analysis of HPB released from human globin have been developed.[45,50] Application of this methodology to snuff dippers demonstrated significantly higher levels of HPB-releasing hemoglobin adducts than in smokers or nonsmokers.[45] This is consistent with the higher levels of urinary total NNAL reported in smokeless tobacco users than in smokers, but further studies are required to pursue this potentially important observation.[33]

Summary

The tobacco-specific nitrosamines NNN and NNK, and the NNK metabolite NNAL, are the strongest carcinogens in smokeless tobacco products. NNN and NNK are considered "carcinogenic to humans" by the International Agency for Research on Cancer. It is widely recognized that tobacco-specific nitrosamines are critical in the etiology of cancer in users of smokeless tobacco products. This recognition has resulted in the FDA's proposal to regulate levels of NNN in smokeless tobacco products, at a level of 1 μg per gram tobacco, on a dry weight basis. Promulgation of this long awaited regulation has the potential to decrease the incidence of frequently fatal cancers of the oral cavity, esophagus, and pancreas in smokeless tobacco users. Urinary biomarkers of exposure to tobacco-specific nitrosamines in smokeless tobacco, such as total NNAL, are now well established and can be used to assess the effects of regulatory strategies on tobacco-specific carcinogen exposure. Other biomarkers, such as carcinogen DNA adducts, are well developed in laboratory animals exposed to tobacco-specific nitrosamines but require further research before they can be widely applied in studies of cancer risk, with the potential to identify those individuals most susceptible to the carcinogenic effects of smokeless tobacco products so that effective preventive measures can be taken.

Acknowledgments

The contributions of all members of the Hecht laboratory to the research described in this chapter and of M. Gerard O'Sullivan and his team for histopathological analysis of tumors induced by tobacco-specific nitrosamines are gratefully acknowledged. This chapter is dedicated to the memory of Dietrich Hoffmann, whose research on the harmful effects of tobacco products, including tobacco-specific nitrosamines, was inspirational. Funding was provided by the National Cancer Institute, mainly through grant CA-81301 and Cancer Center Support Grant 77598.

References

1. International Agency for Research on Cancer. Smokeless tobacco and tobacco-specific nitrosamines. In: *IARC Monographs on the Evaluation of Carcinogenic Risks to Humans*. Vol. 89. Lyon, FR: IARC; 2007:41–417.
2. Stepanov I, Jensen J, Hatsukami D, Hecht SS. New and traditional smokeless tobacco: comparison of toxicant and carcinogen levels. *Nicotine Tob Res*. 2008;10:1773–1782.
3. Richter P, Hodge K, Stanfill S, Zhang L, Watson C. Surveillance of moist snuff total nicotine, pH, moisture, un-ionized nicotine, and tobacco-specific nitrosamine content. *Nicotine Tob Res*. 2008;10(11):1645–1652.
4. Stanfill SB, Connolly GN, Zhang L, et al. Global surveillance of oral tobacco products: total nicotine, unionised nicotine and tobacco-specific *N*-nitrosamines. *Tob Control*. 2011;20(3):e2.
5. Stepanov I, Yershova K, Carmella S, Upadhyaya P, Hecht SS. Levels of (*S*)-*N'*-nitrosonornicotine in U.S. tobacco products. *Nicotine Tob Res*. 2013;15:1305–1310.
6. Lawler TS, Stanfill SB, Zhang L, Ashley DL, Watson CH. Chemical characterization of domestic oral tobacco products: total nicotine, pH, unprotonated nicotine and tobacco-specific *N*-nitrosamines. *Food Chem Toxicol*. 2013;57:380–386.
7. Stepanov I, Gupta PC, Dhumal G, et al. High levels of tobacco-specific *N*-nitrosamines and nicotine in Chaini Khaini, a product marketed as snus. *Tob Control*. 2015;24(e4): e271–274.
8. Stanfill SB, Croucher RE, Gupta PC, et al. Chemical characterization of smokeless tobacco products from South Asia: nicotine, unprotonated nicotine, tobacco-specific *N'*-Nitrosamines, and flavor compounds. *Food Chem Toxicol*. 2018;118:626–634.
9. Magee PN, Barnes JM. The production of malignant primary hepatic tumors in the rat by feeding dimethylnitrosamine. *Brit J Cancer*. 1956;10:114–122.
10. Peto R, Gray R, Brantom P, Grasso P. Effects on 4080 rats of chronic ingestion of *N*-nitrosodiethylamine or *N*-nitrosodimethylamine: a detailed dose-response study. *Cancer Res*. 1991;51(23 Pt 2):6415–6451.
11. Preussmann R, Stewart BW. *N*-nitroso carcinogens. In: Searle CE, ed. *Chemical Carcinogens*. Washington, DC: American Chemical Society; 1984:643–828. ACS Monograph 182. 2nd ed.; Vol. 2.
12. Druckrey H, Preussmann R, Ivankovic S, Schmähl D. Organotrope carcinogen wirkungen bei 65 verschiedenen *N*-nitrosoverbindungen an BD-ratten. *Z Krebsforsch Klin Onkol*. 1967;69:103–201.
13. Lijinsky W. *Chemistry and Biology of* N*-Nitroso Compounds*. Cambridge, England: Cambridge University Press; 1992.

14. Bogovski P, Bogovski S. Animal species in which *N*-nitroso compounds induce cancer. *Int J Cancer*. 1981;27:471–474.
15. Hoffmann D, Hecht SS, Ornaf RM, Wynder EL. *N'*-nitrosonornicotine in tobacco. *Science*. 1974;186:265–267.
16. Hecht SS, Chen CB, Hirota N, Ornaf RM, Tso TC, Hoffmann D. Tobacco-specific nitrosamines: formation from nicotine in vitro and during tobacco curing and carcinogenicity in strain A mice. *J Natl Cancer Inst*. 1978;60(4):819–824.
17. Hecht SS. Biochemistry, biology, and carcinogenicity of tobacco-specific *N*-nitrosamines. *Chem Res Toxicol*. 1998;11:559–603.
18. Stoner GD, Adams C, Kresty LA, et al. Inhibition of *N'*-nitrosonornicotine-induced esophageal tumorigenesis by 3-phenylpropyl isothiocyanate. *Carcinogenesis*. 1998; 19(12):2139–2143.
19. Balbo S, James-Yi S, Johnson CS, et al. (S)-N'-Nitrosonornicotine, a constituent of smokeless tobacco, is a powerful oral cavity carcinogen in rats. *Carcinogenesis*. 2013; 34:2178–2183.
20. Balbo S, Johnson CS, Kovi RC, et al. Carcinogenicity and DNA adduct formation of 4-(methylnitrosamino)-1-(3-pyridyl)-1-butanone and enantiomers of its metabolite 4-(methylnitrosamino)-1-(3-pyridyl)-1-butanol in F-344 rats. *Carcinogenesis*. 2014; 35(12):2798–2806.
21. Andreotti G, Freedman ND, Silverman DT, et al. Tobacco use and cancer risk in the Agricultural Health Study. *Cancer Epidemiol Biomark Prev*. 2017;26(5):769–778.
22. Peterson LA. Context matters: contribution of specific DNA adducts to the genotoxic properties of the tobacco-specific nitrosamine NNK. *Chem Res Toxicol*. 2017;30(1): 420–433.
23. Pegg AE. Multifaceted roles of alkyltransferase and related proteins in DNA repair, DNA damage, resistance to chemotherapy, and research tools. *Chem Res Toxicol*. 2011;24(5):618–639.
24. Hecht SS, Stepanov I, Carmella SG. Exposure and metabolic activation biomarkers of carcinogenic tobacco-specific nitrosamines. *Acc Chem Res*. 2016;49(1):106–114.
25. Yuan JM, Nelson HH, Carmella SG, et al. CYP2A6 genetic polymorphisms and biomarkers of tobacco smoke constituents in relation to risk of lung cancer in the Singapore Chinese Health Study. *Carcinogenesis*. 2017;38(4):411–418.
26. Stepanov I, Hecht SS. Tobacco-specific nitrosamines and their *N*-glucuronides in the urine of smokers and smokeless tobacco users. *Cancer Epidemiol Biomark Prev*. 2005;14: 885–891.
27. Yang J, Carmella SG, Hecht SS. Analysis of *N'*-nitrosonornicotine enantiomers in human urine by chiral stationary phase liquid chromatography-nanoelectrospray ionization-high resolution tandem mass spectrometry. *J Chromatogr B Analyt Technol Biomed Life Sci*. 2017;1044–1045:127–131.
28. Murphy SE, Carmella SG, Idris AM, Hoffmann D. Uptake and metabolism of carcinogenic levels of tobacco-specific nitrosamines by Sudanese snuff dippers. *Cancer Epidemiol Biomark Prev*. 1994;3:423–428.
29. Ahmed HG. Aetiology of oral cancer in the Sudan. *J Oral Maxillofac Res*. 2013;4(2):e3.
30. Kresty LA, Carmella SG, Borukhova A, et al. Metabolites of a tobacco-specific nitrosamine, 4-(methylnitrosamino)-1-(3-pyridyl)-1-butanone (NNK), in the urine of smokeless tobacco users: relationship of urinary biomarkers and oral leukoplakia. *Cancer Epidemiol Biomark Prev*. 1996;5:521–525.
31. Hecht SS, Carmella SG, Ye M, et al. Quantitation of metabolites of 4-(methylnitrosamino)-1-(3-pyridyl)-1-butanone after cessation of smokeless tobacco use. *Cancer Res*. 2002;62:129–134.

32. Hecht SS, Carmella SG, Murphy SE, et al. Similar exposure to a tobacco-specific carcinogen in smokeless tobacco users and cigarette smokers. *Cancer Epidemiol Biomark Prev*. 2007;16(8):1567–1572.
33. Rostron BL, Chang CM, van Bemmel DM, Xia Y, Blount BC. Nicotine and toxicant exposure among U.S. smokeless tobacco users: results from 1999 to 2012 National Health and Nutrition Examination Survey data. *Cancer Epidemiol Biomark Prev*. 2015; 24(12):1829–1837.
34. Hecht SS, Carmella SG, Stepanov I, Jensen J, Anderson A, Hatsukami DK. Metabolism of the tobacco-specific carcinogen 4-(methylnitrosamino)-1-(3-pyridyl)-1-butanone to its biomarker total NNAL in smokeless tobacco users. *Cancer Epidemiol Biomark Prev*. 2008;17(3):732–735.
35. Hatsukami DK, Stepanov I, Severson H, et al. Evidence supporting product standards for carcinogens in smokeless tobacco products. *Cancer Prev Res (Phila)*. 2015;8(1): 20–26.
36. Benowitz NL, Renner CC, Lanier AP, et al. Exposure to nicotine and carcinogens among Southwestern Alaskan Native cigarette smokers and smokeless tobacco users. *Cancer Epidemiol Biomark Prev*. 2012;21(6):934–942.
37. Yang J, Villalta PW, Upadhyaya P, Hecht SS. Analysis of O^6-[4-(3-pyridyl)-4-oxobut-1-yl]-2'-deoxyguanosine and other DNA adducts in rats treated with enantiomeric or racemic *N*'-nitrosnornicotine. *Chem Res Toxicol*. 2015;29:87–95.
38. Zhao L, Balbo S, Wang M, et al. Quantitation of pyridyloxobutyl-DNA adducts in tissues of rats treated chronically with (*R*)- or (*S*)-*N*'-nitrosonornicotine (NNN) in a carcinogenicity study. *Chem Res Toxicol*. 2013;26(10):1526–1535.
39. Zhang S, Wang M, Villalta PW, Lindgren BR, Lao Y, Hecht SS. Quantitation of pyridyloxobutyl DNA adducts in nasal and oral mucosa of rats treated chronically with enantiomers of *N*'-nitrosonornicotine. *Chem Res Toxicol*. 2009;22:949–956.
40. Upadhyaya P, Kalscheuer S, Hochalter B, Villalta PW, Hecht SS. Quantitation of pyridylhydroxybutyl-DNA adducts in liver and lung of F-344 rats treated with 4-(methylnitrosamino)-1-(3-pyridyl)-1-butanone and enantiomers of its metabolite 4-(methylnitrosamino)-1-(3-pyridyl)-1-butanol. *Chem Res Toxicol*. 2008;21:1468–1476.
41. Murphy SE, Palomino A, Hecht SS, Hoffmann D. Dose-response study of DNA and hemoglobin adduct formation by 4-(methylnitrosamino)-1-(3-pyridyl)-1-butanone in F344 rats. *Cancer Res*. 1990;50:5446–5452.
42. Stepanov I, Muzic J, Lee CT, et al. Analysis of 4-hydroxy-1-(3-pyridyl)-1-butanone (HPB)-releasing DNA adducts in human exfoliated oral mucosa cells by liquid chromatography-electrospray ionization-tandem mass spectrometry. *Chem Res Toxicol*. 2013;26:37–45.
43. Khariwala SS, Ma B, Ruszczak C, et al. High level of tobacco carcinogen-derived DNA damage in oral cells is an independent predictor of oral/head and neck cancer risk in smokers. *Cancer Prev Res (Phila)*. 2017;10(9):507–513.
44. Ma B, Ruszczak C, Jain V, et al. Optimized liquid chromatography nanoelectrospray-high-resolution tandem mass spectrometry method for the analysis of 4-hydroxy-1-(3-pyridyl)-1-butanone-releasing DNA adducts in human oral cells. *Chem Res Toxicol*. 2016;29(11):1849–1856.
45. Carmella SG, Kagan SS, Kagan M, et al. Mass spectrometric analysis of tobacco-specific nitrosamine hemoglobin adducts in snuff dippers, smokers, and non-smokers. *Cancer Res*. 1990;50:5438–5445.
46. Ehrenberg L, Osterman-Golkar S. Alkylation of macromolecules for detecting mutagenic agents. *Teratog Carcinog Mutagen*. 1980;1(1):105–127.
47. Carmella SG, Hecht SS. Formation of hemoglobin adducts upon treatment of F344 rats with the tobacco-specific nitrosamines 4-(methylnitrosamino)-1-(3-pyridyl)-1-butanone and *N*'-nitrosonornicotine. *Cancer Res*. 1987;47:2626–2630.

48. Carmella SG, Kagan SS, Hecht SS. Evidence that a hemoglobin adduct of 4-(methylnitrosamino)-1-(3-pyridyl)-1-butanone is a 4-(3-pyridyl)-4-oxobutyl carboxylic-acid ester. *Chem Res Toxicol.* 1992;5(1):76–80.
49. Hecht SS, Trushin N, Rigotty J, et al. Complete inhibition of 4-(methylnitrosamino)-1-(3-pyridyl)-1-butanone induced rat lung tumorigenesis and favorable modification of biomarkers by phenethyl isothiocyanate. *Cancer Epidemiol Biomark Prev.* 1996;5: 645–652.
50. Hecht SS, Carmella SG, Murphy SE. Tobacco-specific nitrosamine-hemoglobin adducts. *Methods Enzymol.* 1994;231:657–667.
51. Ma B, Zarth AT, Carlson ES, et al. Identification of more than 100 structurally unique DNA-phosphate adducts formed during rat lung carcinogenesis by the tobacco-specific nitrosamine 4-(methylnitrosamino)-1-(3-pyridyl)-1-butanone. *Carcinogenesis.* 2018; 39(2):232–241.
52. Carlson ES, Upadhyaya P, Villalta PW, Ma B, Hecht SS. Analysis and identification of 2'-deoxyadenosine-derived adducts in lung and liver DNA of F-344 rats treated with the tobacco-specific carcinogen 4-(methylnitrosamino)-1-(3-pyridyl)-1-butanone and enantiomers of its metabolite 4-(methylnitrosamino)-1-(3-pyridyl)-1-butanol. *Chem Res Toxicol.* 2018;31(5):358–370.
53. Lao Y, Villalta PW, Sturla SJ, Wang M, Hecht SS. Quantitation of pyridyl-oxobutyl DNA adducts of tobacco-specific nitrosamines in rat tissue DNA by high performance liquid chromatography-electrospray ionization-tandem mass spectrometry. *Chem Res Toxicol.* 2006;19:674–682.
54. Ma B, Villalta PW, Zarth A, et al. Comprehensive high resolution mass spectrometric analysis of DNA phosphate adducts formed by the tobacco-specific lung carcinogen 4-(methylnitrosamino)-1-(3-pyridyl)-1-butanone (NNK). *Chem Res Toxicol.* 2015;28: 2151–2159.
55. Zarth AT, Upadhyaya P, Yang J, Hecht SS. DNA adduct formation from metabolic 5'-hydroxylation of the tobacco-specific carcinogen *N'*-nitrosonornicotine in human enzyme systems and in rats. *Chem Res Toxicol.* 2016;29(3):380–389.

CHAPTER EIGHT

Reducing carcinogen levels in smokeless tobacco products

Stephen B. Stanfill
U.S. Centers for Disease Control and Prevention, Atlanta, GA, United States

Introduction

Newly planted tobacco generally contains low levels of carcinogenic agents, whereas certain smokeless tobacco (ST) products may contain high levels. The problematic content results from agents naturally present or introduced during cultivation and processing, and from the interactions of these agents.[1–8] The agents in the final product may result from accumulation of soil metals and nitrate, plant biosynthesis of alkaloids, byproducts of microbial activity, residues from fire-curing, or additives such as alkaline agents to boost pH, which increases nicotine absorption, and the inclusion of areca nut, which is carcinogenic. Key processes in cultivation and processing are discussed, especially those that contribute most to toxicity, including the role of the tobacco-growing environment, soil constituents, and other agronomic parameters.[2–8] The agents discussed are toxic metals, tobacco-specific nitrosamines (TSNA), other nitrosamines, mycotoxins, smoke-related compounds, polycyclic aromatic hydrocarbons (PAH) resulting from fire-curing, and ingredients.[5,6] The presence of fungi can lead to mycotoxin formation in ST products.[9,10] Microbial mechanisms of nitrite generation and excretion that initiate the formation of carcinogenic TSNA and the conditions under which these processes occur are discussed. Researchers have identified microorganisms that may drive these processes.[2,10,11] It has also been shown in snus that modifying certain processes and implementing ongoing monitoring and adherence to allowable levels of these agents or their precursors lowers the levels of carcinogens.[12] Potential means of detecting and eliminating microorganisms in tobacco leaf are presented.

Tobacco product toxicity has been studied extensively for decades with agents such as toxic metals and TSNA commonly cited in ST.[3,5–7] Metals can be absorbed and enter tobacco leaves. Researchers have found that

Smokeless Tobacco Products
ISBN: 978-0-12-818158-4
https://doi.org/10.1016/B978-0-12-818158-4.00008-X

soil concentrations of cadmium, lead, zinc, and copper are associated with accumulation in burley, Virginia, and oriental tobacco, with highest concentrations in the leaves on lower parts of the plant.[4] The presence of soil contamination of metals such as lead, nickel, and cadmium can also affect the levels in tobacco.[13] Fermented products, including moist snuff, dry snuff, and khaini, contain higher levels of TSNA than pasteurized products, such as snus.[5,14–18] Sudanese toombak has among the highest TSNA concentrations.[1,16] Tobacco leaves at harvest contain almost no TSNA,[1] but finished products have widely varying concentrations of individual TSNA compounds. Snus, a pasteurized product common in Sweden, often has parts per billion (ng/g) levels,[16,19] while khaini and US-made moist and dry snuff are often in the parts per million (μg/g) range.[14–16] Surprisingly, toombak has TSNA in the parts per thousand (mg/g) levels.[1] The causes of this extreme variation are beginning to be understood and are addressed in this chapter.

Table 8.1 shows where key toxicants and carcinogens come from.

Factors that influence carcinogens and their precursors: soil, curing, microbial activity, and additives

Table 8.2 illustrates how harmful characteristics come to be present in ST, from preplanting to consumption. Measurement of constituents has defined harmful agents and their approximate levels in a variety of smokeless products. The concentrations of toxicants in a wide spectrum of products available worldwide are recorded in the literature.[5] IARC Group I carcinogens[20] in tobacco products can include mycotoxins, PAH, volatile aldehydes, areca nut, and TSNA such as N-nitrosonornicotine (NNN).[3,5,10,14–16,21,22] There are many factors that influence the levels of TSNA in ST.[2,8,12] A potential framework for the implementation of options available to manufacturers to reduce NNN levels is described in the proposed US Food and Drug Administration (FDA) rule.[23] Tobacco contains the IARC Group 1 carcinogenic metals arsenic, beryllium, cadmium, nickel compounds, and the radioisotope polonium-210. Metals such as aluminum, chromium, and nickel cause sensitization, whereas, barium and mercury are irritants. Metal content is highly influenced by soil and environment.[3,4,13]

Tobacco species used in ST products (mostly *Nicotiana tabacum* and *N. rustica*) are members of the Solanaceae family, and each contains the genetically influenced alkaloids nicotine, nornicotine, anabasine, and

Table 8.1 Sources of key toxicants and carcinogens that accumulate or form in smokeless tobacco.

Soil absorption or leaf deposition	Leaf deposition during fire-curing	Microorganisms: fungi	Microorganisms: Bacteria or fungi	Additives
Metals	**Polycyclic aromatic hydrocarbons (PAH)**	**Mycotoxins**	**Microbial byproduct**	**Nontobacco plant material**
Group 1	Group 1	Group 1	*Nitrosation agent*	Group 1
Arsenic	Benzo[*a*]pyrene	Aflatoxins (mixtures of)	Group 2B	Areca nut
Beryllium	Group 2A	Group 2B	Nitrite	
Cadmium	Dibenz[*a,h*]anthracene	Aflatoxin M1	**Reaction products of nitrite with tobacco alkaloids**	Liver toxicants
Chromium (VI)	Group 2B	Ochratoxin A	*Tobacco-specific N-nitrosamines (TSNA)*	Tonka bean
Nickel compounds	Benz[*a*]anthracene	Sterigmatocystin	Group 1	
Polonium-210 and their related compounds	Benzo[*b*]fluoranthene		N-nitrosonornicotine (NNN)	Stimulants
Group 2A	Benzo[*j*]fluoranthene		4-(methylnitrosamino)-1-(3-pyridyl)-1-butanone (NNK)	Khat
Inorganic lead compounds	Benzo[*k*]fluoranthene		4-(methylnitrosamino)-1-(3-pyridyl)-1-butanol (NNAL)	Caffeine
Group 2B	Dibenzo[*a,i*]pyrene		**Reaction products of nitrite with certain secondary or tertiary amines**	
Cobalt	Indeno[1,2,3-*cd*]pyrene		*Volatile N-nitrosamines (VNA)*	
Sensitization	5-Methylchrysene		Group 2A	
Aluminum	Naphthalene		N-nitrosodimethylamine (NDMA)	
Chromium	**Volatile aldehydes (VOC)**			
Cobalt				
Nickel				
Irritants	Group 1			

(Continued)

Table 8.1 Sources of key toxicants and carcinogens that accumulate or form in smokeless tobacco.—cont'd

Soil absorption or leaf deposition	Leaf deposition during fire-curing	Microorganisms: fungi	Microorganisms: Bacteria or fungi	Additives
Barium	Formaldehyde		Group 2B	
Mercury	Group 2B		N-nitrosopyrrolidine (NPYR)	
Inorganic nitrogen	Acetaldehyde		N-nitrosopiperidine (NPIP)	
Group 2B			N-nitrosomorpholine (NMOR)	
Nitrate			N-nitrosodiethanolamine (NDEA)	
			Nitrosoacids (NA)	
			Group 2B	
			N-nitrososarcosine (NSAR)	
			Reaction products of ethanol and urea	
			Carbamates	
			Group 2A	
			Ethyl carbamate	

Carcinogen Group is based on the International Agency for Research on Cancer (IARC) scheme.

Table 8.2 How harmful characteristics come to be present in smokeless tobacco products.

Cultivation	Harvest	Curing
• Choice of tobacco type or species affects the alkaloid and sugar content but also taste/aroma. • Plant uptake of nitrate and soil metals occurs. • Tobacco synthesizes nicotine, other alkaloids, and reactive amines that are addictive, toxic, or can react to form TSNA and other nitrosamines. • Soil, soil metals, microorganisms, and ag-chemicals (pesticides) may deposit on the leaves.	• Contamination with soil constituents, including soil microbes, may occur if leaves are laid on or remain in contact with the soil. • If tobacco is harvested with bare hands, transfer of dermal-associated microbes is possible.	**Air curing** • Tobacco cells dry and rupture. This releases nitrate that becomes available for microbes. • With poor ventilation, fungi may grow and produce mycotoxins. **Fire curing** • Tobacco cells dry and rupture. This releases nitrate that becomes available for microbes. • Smoke-related chemicals (PAH, VOC) can deposit on leaves. • Certain gases (NO_x) can produced and react with alkaloids to form TSNA.
Agents accumulated in or on the leaf: nicotine and minor alkaloids (TSNA precursors), reactive amines (nitrosamine and nitro acid precursors), nitrate, toxic soil metals, pesticides, and microorganisms (bacteria, fungi, viruses)	Agents accumulated in or on the leaf: soil particles, soil metals, agricultural chemicals, microorganisms (from the soil and human skin)	Agents that may be accumulated or formed: PAH, VOC, mycotoxins, nitrite, TSNA

(Continued)

Table 8.2 How harmful characteristics come to be present in smokeless tobacco products.—cont'd

Fermentation/Aging/Long-term storage	Formulation	Product use
• These processes are often characterized by rapid microbial proliferation. • Anaerobic environments can promote conversion of nitrate to nitrite by respiratory nitrate reductase and associated transporters. • TSNA and other nitrosamines may be formed when nitrite reacts with alkaloids or reactive amines in the process of nitrosation.	• Additives may affect addiction liability or toxicity. The use of pH-boosting agents may enhance nicotine absorption. • Areca nut is a known carcinogen. • If a product is hand-mixed, dermal-microbes may be introduced.	• ST introduces harmful and addictive chemicals into the oral cavity, where they can be absorbed. • Addition of areca nut can cause oral submucosal fibrosis. • ST may introduce live microorganisms that adversely affect health. Transfer of pathogens and opportunistic organisms into the body is possible. • ST is associated with cancer, heart disease, and other disorders.
Agents that may be formed: nitrite, TSNA, other nitrosamines	Agents added: pH-boosters (increase pH and percent of free nicotine), areca nut (carcinogen), khat (psychoactive), tonka bean	

anatabine.[1,7,24,25] Another species, *N. glauca*, lacks nicotine but is toxic and sometimes lethal because of high concentrations of anatabine.[26] Tobacco alkaloids contribute to the formation of TSNA; nicotine and nornicotine are precursors to the most potent human carcinogens, NNN and 4-(methylnitrosamino)-1-(3-pyridyl)-1-butanone (NNK).[7,8,27] The choice of species and varieties is an agronomic decision based on chemical characteristics and the product for which it is to be used.[7] In particular, the use of *N. rustica*, which has high concentrations of nicotine and nornicotine, contribute to high concentrations of NNN and NNK.[1,16,28] Plant growth and the concentrations of TSNA precursors (nitrate, alkaloids) and TSNA are increased by nitrate fertilizer.[7,22,29,30]

Tobacco accumulates inorganic agents from the soil, including metals necessary for growth and toxic ones.[3,4,7,9] In a review of studies of a limited number of ST products, those from Pakistan, India, and Ghana had higher levels of metals than those from Canada, the United States, and Sweden. Lead, nickel, and chromium were higher in products from Pakistan and lower in products from India. High copper was found in products from India and Ghana.[3] Metals that are not plant macro- or micronutrients likely are absorbed from soil or soil amendments[13] or deposition on tobacco leaves.[31] Monitoring the metal content in tobacco fields is a prudent step to minimize the presence of these metals in ST.[3]

Microbial and chemical constituents are affected at various steps in tobacco processing.[2,21,28,32,33] During cultivation, inorganic substances (nitrate and metals) are absorbed from the soil. Organic compounds (nicotine, minor alkaloids, and other reactive amines) are synthesized[7,29] and microorganisms and other materials deposit on the leaves.[34,35] Three main components lay the groundwork for TSNA formation: absorbed soil nitrate, nitrite-generating organisms, and alkaloids.[2,8,12] Higher temperatures and humidity during cultivation and other processes also favor TSNA generation.[2,36–38]

At time of harvest, leaves may be picked individually or the entire stalk may be harvested then cured.[7] For the tobacco used in toombak, stalks are cut and sometimes laid on the ground for 45 days, allowing microorganisms and other materials to be deposited.[28] Because tobacco is not generally washed at harvest, leaf surfaces may contain soil constituents, soil metals, agricultural chemical residues, and live or deceased bacteria and fungi. Washing or disinfecting at harvest would remove at least some of the microbes and other substances.[13,34,35,39]

After harvest, drying of tobacco is accomplished using simple forms of curing (sun-curing, air-curing) or more elaborate flue-curing or fire-curing. In presence of high humidity (70%), temperatures ranging from 10 to 32°C, and poor ventilation, fungi can produce mycotoxins such as aflatoxins and ochratoxins.[9] Fire-curing, which involves exposure to smoke from burning wood and sawdust,[7] can increase levels of the PAH benzo[a]pyrene and formaldehyde,[19] both Group 1 IARC carcinogens. Concentrations of the PAH phenanthrene, fluoranthene, pyrene, benzo[a]anthracene, chrysene, and benzo[a]pyrene in fire-cured tobacco are higher than in air-cured.[40] Also, NO_x gases present during curing can react with alkaloids to form TSNA.[30] As drying occurs, tobacco cells rupture and cell constituents (e.g., sugars, nitrate, and alkaloids) become available for biotransformation if microbes are present.[2,7,8,12]

Microbes may enter via agricultural and process environments and also human contact.[2,11,33,37,41] Molecular studies have shown that tobacco and its products vary in their microbial communities.[2,11,33,41] Moist snuff showed low diversity and contained mainly Firmicutes. Dry snuff contained a wide array, mostly Firmicutes, Proteobacteria, and to a lesser extent Actinobacteria.[11,33] Toombak, a highly fermented Sudanese product with the highest known TSNA concentration,[1] contained Actinobacteria and less abundant Firmicutes.[11,33] Some Actinobacterial species (*Corynebacterium* and *Enteractinococcus*) capable of generating nitrite found in tobacco are alkaline- and thermotolerant and may remain viable through late-stage fermentation, when higher pH and temperature exist.[2,11,33,42] *Corynebacterium* species such as *C. stationis*, *C. ammoniagenes*, and *C. casei* reduce nitrate to nitrite[2,42] and are present in fermented cigar tobacco[2] and toombak. Some *Staphylococcus* may also produce nitrite.[33]

During fermentation and aging, oxygen levels may be lower.[2,36,37] In low oxygen environments, microbes can perform respiratory nitrate reduction—where nitrate instead of oxygen is used to maintain respiration. Once nitrite is generated, it is excreted by the microbial transporters because of its cytotoxicity.[43,44] Aging and fermentation are active metabolic systems in which microbes import, metabolize, and excrete compounds. If excreted nitrite accumulates, it reacts with tobacco-associated alkaloids to form TSNA or with secondary or tertiary amines that form volatile nitrosamines and nitrosoacids.[2,8,32,36] Toombak especially, which is made with lengthy processes under anaerobic conditions,[1,28] may allow more time for nitrite to be generated and react to form nitrosamines. This and other factors help to explain the wide range of TSNA concentrations found in products.

ST production is influenced by physicochemical conditions (temperature, pH, moisture) and the extent of various biochemical and chemical actions. Products tend to become more toxic in steps from the growing tobacco plant to the finished form if counteractive steps are not taken.[2,4,8,12,21,32,36] Identifying specific microbes, their genes, and the mechanisms[11,33,38] that form nitrite is important in decreasing TSNA formation. Research on tobacco, moist snuff, dry snuff, and toombak has allowed the identification of genes in nitrate metabolism.[33,38] Pasteurization eliminates microorganisms, including those with nitrite-generating capabilities, and results in reduced concentrations of carcinogenic TSNA.[12]

Fig. 8.1 shows the nitrogen reduction pathways for the production of TSNA and summarizes the problem and possible solutions. Starting in curing, nitrosamine formation involves two steps, the microbial generation and excretion of nitrite, and the chemical reaction that generates nitroso compounds.[2,8,12,32,36] Microbial generation and excretion of nitrite is mediated by respiratory nitrate reductases and ion transporters that export nitrite. Respiratory nitrate reductases are expressed under low-oxygen conditions.[43,44] This is followed by oxygen-independent nitrosation in which nitrite reacts with amines (including tobacco alkaloids) to form TSNA, volatile nitrosamines, and nitrosoacids.[8,27]

The amount of nitrite generated influences TSNA formation throughout production.[2,8,32,36] Aside from the chemical activity of these organisms, the introduction and retention of viable microorganisms in the oral cavity is a concern.[45,46] Processes to decrease or eliminate nitrite-excreting organisms have been shown to decrease TSNA levels.[12,32,39]

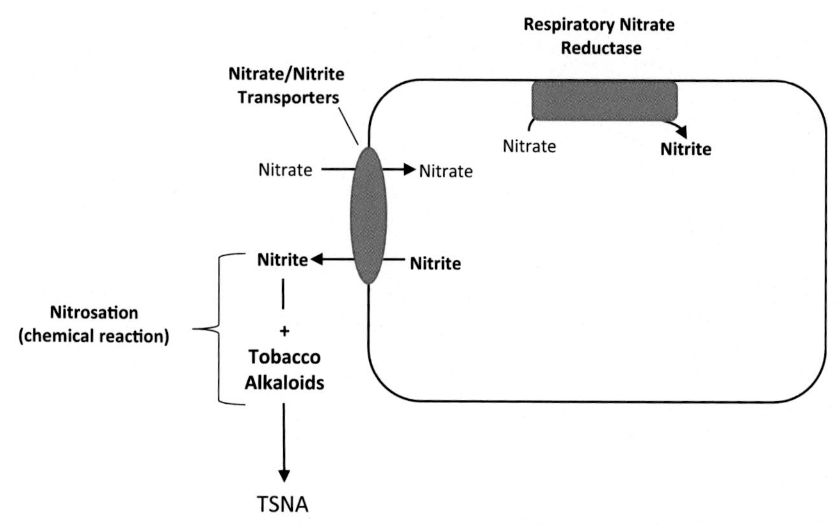

Figure 8.1 Microbial and chemical steps that may form TSNA.

The presence of nicotine promotes the repeated use of tobacco products and on-going exposure to harmful agents. The amount of nicotine absorbed is influenced by total nicotine and pH,[47,48] the latter manipulated by additives during formulation. Increasing the pH of ST increases the percentage of nicotine that is converted to the readily absorbable unprotonated (free) form.[49] Because unprocessed cured tobacco is generally acidic (pH 5 to 6.5),[14] products with higher pH levels have generally been augmented with alkaline agents such as sodium carbonate, sodium bicarbonate, ammonium carbonate, calcium hydroxide, slaked lime, and ashes from burned wood, plants, and fungi. Absorbed nicotine contributes to dependency and cardiovascular disease.[5,6] The inclusion of areca nut, the fruit of the areca palm (*Areca catechu*), in products is of concern because it is an IARC Group 1 carcinogen.[5,22]

Techniques for detecting causal agents

Table 8.3 shows techniques for assessing microbial contamination.

Analysis of ST products vary but depend heavily on gas chromatography (GC) and liquid chromatography (LC) linked to various types of mass spectrometry (MS) or other detectors, including flame ionization detectors (FID) and inductively coupled plasma (ICP). Nicotine content can be assessed by GC/FID[49] or GC/MS.[19,56] Volatile organic compounds (VOC) and PAH[19] can be measured with GC/MS or GC/MS/MS. Minor alkaloids have recently been measured with GC/MS/MS.[25] Many of the toxic metals are quantified by high pressure liquid chromatography (HPLC) coupled to an ICP detector.[57] TSNA,[14,19] aflatoxins,[10] and arecoline and other areca constituents (areca nut)[58] can be measured with LC/MS/MS. Other methods have been used and could be substituted with appropriate validation.

Where funding or space may be limited, portable instrumentation such as GC/MS offers a means of detecting and quantifying alkaloids (both nicotine and minor alkaloids), nontobacco constituents, and flavor-related compounds.[17,25,56,58,59] The amount of nicotine as free nicotine is calculated using pH.[14] X-ray fluorescence spectroscopy can detect metals and alkaline agents[60]; a hand-held unit is available. Infrared spectroscopy can identify tobacco species (*N. tabacum, N. rustica, N. glauca*), nontobacco plant materials (areca nut, tonka bean), and alkaline agents (magnesium carbonate, slaked lime).[16,17,61] Infrared spectroscopy is now available as a portable scanner.[62] Methods to measure ammonia, nitrate, and nitrite[19] are available. Culture

Table 8.3 Techniques for assessing microbial contamination in tobacco and tobacco products.

Culture counts using standard culture media	Standard culture media such as tryptic soy agar (TSA), MacConkey agar plates (MAC), mannitol salt agar (MSA), and sheep blood agar[11,37] are used to assess tobacco microbes. Sabouraud's dextrose agar is used to assess fungal contamination.[9] A single culture medium will not capture diversity in ST products. On a given type of media, certain microbial taxa may be overrepresented and less abundant taxa may not grow and thus remain unidentified.[50]	This laborious approach requires sterile facilities, supplies, and microbiology expertise and may be cost prohibitive in some cases. Less expensive approaches may have advantages.
Culture counts using disposable microbiological films	A disposable product contains a lower film that is inoculated and an attached upper film that is lowered to cover it. The inoculated film is then incubated and read for colony counts. This method can be used for aerobic bacteria, coliform bacteria, Enterobacteriaceae, *E. coli*, heterotrophic lactic acid bacteria, yeast, and mold.[51]	Film-based medium is relatively inexpensive and does not require the pouring of media plates. This approach has been used successfully for 30 years in the food industry.
Detection of cell viability by handheld luminometer	Handheld luminometers, which test for the presence of ATP, have been compared by several manufacturers.[52–54] Devices exist that can yield total visible count for specific bacterial groups (Enterobacteriaceae, coliform, *E. coli*, and *Listeria* spp.)[55]	Measurements can be made in 5–20 s. These devices are used in the healthcare, pharmaceuticals, water treatment, and food-related industries and could be applied to tobacco products, preparations, or ingredients on-site.
Detection of organisms by qPCR	qPCR (quantitative polymerase chain reaction) is a technique in which probes against specific gene regions bind, replicate, and allow quantification. Test kits exist for various species.	Organisms for which probes are not available can be detected by adapting those for nitrate-reducing organisms and specific genes (e.g., *narG*, *narK*, *narT*, and *napA*) from key genera (*Corynebacterium*, *Staphylococcus*, etc.)[33,38]

plates specific to bacteria[37] and fungi[9] may be helpful for identifying contaminated products.

Strategies for reducing causal agent levels

Table 8.4 illustrates how harmful characteristics can be reduced in ST, from preplanting to consumption.

Plant genetic and molecular approaches to lower TSNA precursors

Molecular and nonmolecular approaches for producing tobaccos with lower TSNA precursors (nitrate, alkaloids) have been reported.[63] Nitrate that is converted by microorganisms to nitrite then reacts with nornicotine; these two precursors form the TSNA carcinogen NNN. Tobacco that expresses a constitutively active nitrate reductase (NR) have decreased free nitrate levels in the growing leaves, which results in drastically lower TSNA levels in the cured leaves.[64] Another means of decreasing NNN is to halt the activity of the enzyme nicotine demethylase, which converts nicotine to nornicotine in the plant. Varieties with lower nornicotine are preferable to those with higher concentrations. Mutation breeding related to nicotine N-demethylase has been used instead of genetic modification for producing nornicotine-free tobacco.[65] Also, the insertion of a sequence that induces RNA interference-induced suppression of nicotine demethylase activity decreases the level of nornicotine in the cured leaf so that less NNN is produced. A gene with a sequence complementary to the mRNA encoding nicotine demethylase essentially blocks the production of that protein so that nicotine is not converted to nornicotine.[63] Morita and colleagues found that blocking or eliminating the gene for the jasmonate-inducible alkaloid transporter (Nt-JAT1), which is involved in alkaloid translocation from the root to the leaves, produces leaves free of alkaloids.[66] Targeting genes encoding enzymes necessary for alkaloid synthesis[63,67] could also produce a tobacco with little or no alkaloid content. In any case, a tobacco without nicotine and nornicotine to react with nitrite would likely produce very little NNN, NNK, and NNAL (an oxidation product of NNK) during processing.

Disinfection of tobacco

Disinfection of tobacco leaf surface with a solution such as dilute bleach (hypochlorite and water) removes microbes[39] and also removes soil that may contain metals and agricultural chemicals.[4,34,35] Protocols used to disinfect

Table 8.4 How harmful characteristics can be remediated in smokeless tobacco products.

Cultivation	Harvest	Curing
• Avoid planting tobacco in soils with high levels of toxic metals. • Use *Nicotiana tabacum* varieties, including those that are genetically modified, that accumulate lower levels of nitrate and alkaloid precursors. • Avoid planting *Nicotiana rustica* or other tobaccos with high levels of TSNA precursors (especially nicotine and nornicotine). • Avoid planting *Nicotiana glauca*, which has high levels of the toxicant anatabine. • Avoid excessive application of agricultural-chemicals such as pesticides and nitrate fertilizer, especially late in the season.	• Avoid soil and direct dermal contact with tobacco. Lay harvested tobacco on a barrier such as a plastic sheet. • Wash and disinfectant prior to curing to decrease the amount of soil, soil metals, soil constituents (manure), agricultural chemical residues, and microorganisms deposited on the leaves. • Allow leaves to fully dry after washing and prior to curing to prevent fungal growth.	**Air-curing** • Maintain proper ventilation and clean conditions free from dust, fungi, and insects. • Prevent temperature and humidity conditions conducive to mold growth. • Prevent accumulation of NO_x gases that can form TSNA. **Fire-curing** • Minimize or eliminate fire-curing. • Consider using smoke flavoring as an additive in lieu of fire-curing.
Agents that may be eliminated or levels minimized: nicotine and minor alkaloids (TSNA precursors), soil metals, pesticides, microorganisms	Agents that may be eliminated or levels minimized: soil, soil metals, soil constituents (manure), agricultural chemicals, microorganisms (soil, dermal)	Agents that may be eliminated or levels minimized: PAH, VOC, mycotoxins, NO_x gases, nitrite, TSNA
Fermentation/Aging/Long-term storage	**Formulation**	**Product sale and use**
• Omit or shorten fermentation, aging, and storage, if possible. • Use heat treatment or electron beam irradiation. • If fermentation is used, clean vats between batches.	• Avoid areca nut. • Eliminate or decrease the use of pH-boosting agents (various carbonates, calcium hydroxide, slaked lime, and ashes from wood, plants, and fungi),	• Refrigerate products during transport and presale storage. • Encourage consumers to refrigerate or store products at cooler temperatures.

(*Continued*)

Table 8.4 How harmful characteristics can be remediated in smokeless tobacco products.—cont'd

Fermentation/Aging/Long-term storage	Formulation	Product sale and use
• If fermentation is used, add only fermentation organisms that do not generate nitrite. • Investigate the use of microbes that accumulate nitrite or convert it to less harmful forms such as N_2 gas by denitrification. • Monitor and control TSNA levels throughout processing.	which make nicotine more easily absorbed. • Use humectants (glycerin, propylene, glycol) that have low water activity and inhibit microbes. • Add nitrite-scavenging agents (vitamin C, green tea extract, etc.) that trap nitrite and prevent it from reacting to form nitrosamines. • Avoid hand mixing to prevent dermal contact. • Use food grade additives that are free of microbial contamination.	
Agents that may be eliminated or levels minimized: nitrite, TSNA, other nitrosamines	Agents that may be eliminated or levels minimized: pH-boosters, areca nut (carcinogen), khat (psychoactive)	Agents that may be eliminated or levels minimized: continued generation of TSNA and other nitrosamines

food products such as leafy greens may prove valuable to tobacco manufacturers. While only applicable to food, the use of disinfection, gloves, and monitoring are covered in the FDA Food Safety Modernization Act (FSMA), signed into law in January 2011.[68] The processing of snus, which is similar to the processing of food, has led to reproducibly lower toxicant levels, as seen in Swedish-made products under the GothiaTek standard.[12]

Pasteurization of snus

Pasteurized products have lower TSNA levels than fermented products.[5,12,14,19] Inside a closed process blender, water and sodium chloride are blended with ground tobacco (either air- or sun-cured). Hot water and injected steam increase the temperature of the mixture to 80—100°C for several hours.[12] The material is then cooled prior to the addition of flavors and humectants. The finished product is fed to an automatic packer. Although differences exist between products made by different manufacturers, the levels of TSNA are extremely low.[5]

Electron beam technology

High-energy electron beam (eBeam) uses a compact linear accelerator to generate highly energetic (10 MeV) electrons with regular electricity. It is approved for pasteurizing foods and sterilizing medical devices. It is called cold pasteurization because it irradiates without generating excess heat that might cause undesirable product changes.[69] The use of eBeam could eliminate organisms that generate nitrite and thus the formation of TSNA and other nitrosamines. It might also eliminate other potentially viable harmful organisms.

Microwave technology

Microwave energy has been used for pasteurizing, sterilizing, bacterial destruction, and enzyme deactivation in food processing and nutraceutical and pharmaceutical production. When exposed to microwave radiation, water molecules within a material absorb the energy and internal heat is generated by molecular vibrations. Continuous microwave heating produces denaturation of critical biomolecules, reduced cell efficacy, and often cell death.[70—72] This technology could be applied to tobacco products.

Cleaning fermentation equipment

Fermentation is common in the preparation of tobacco products, including cigars[2] and ST[7,37] It lasts a few weeks and is characterized by chemical and microbial changes.[2,28,32] The elimination of nitrate-reducing microbes active during fermentation is important.[2,32] Fisher and other researchers described cleaning of fermentation vats between batches and spiking of vats with nonnitrate reducing bacteria as an effective means of lowering TSNA.[32]

Nitrite scavengers

Completely eliminating nitrite-generating organisms is the most straightforward means of minimizing nitrite. Failing that, nitrite scavenging agents can be used. Addition of agents such as tocopherol, vitamin C, morpholine, polyphenols, and the green tea component epigallocatechin gallate are examples.[12,73,74]

Conclusion

This chapter highlights the presence of certain toxicants and carcinogens in ST, including toxic metals, aflatoxins, TSNA, other nitrosamines, volatile aldehydes, PAH, and areca nut. The growing environment and soil are the source of toxic metals. Because carcinogenic metals such as arsenic, beryllium, cadmium, nickel, and polonium-210 are not essential nutrients for tobacco, they are unlikely to be absorbed but result rather from the deposition of contaminated soil onto the leaves.[3,4] Agricultural chemicals and nitrate-containing fertilizer are added during cultivation.[7] PAH and volatile aldehydes may deposit on the leaf surface during fire-curing.[19] Harvested tobacco contains agricultural chemical residues on the leaf surface, whereas nitrate remains in the tissue.[12,34,35]

Aflatoxins and ochratoxins are the products of fungi such as *Aspergillus* and *Penicillium*.[75] Mold growth is favored by high humidity, temperatures of 10–32°C, and poor ventilation during curing. Mold has been reported for products made in Pakistan.[9] Aflatoxins are present in dry snuff products made in the United States.[10] During curing as tobacco tissue dries, cells rupture and nitrate begins to be converted to nitrite.[8,12] Certain bacteria contain respiratory nitrate reductases and transporters that facilitate the conversion of nitrate to nitrite and its export from the bacterial cell.[2,33,38] This respiratory process is active in low oxygen environments when microbes use nitrate in lieu of oxygen to maintain respiration.[43] Low oxygen may exist during fermentation, aging, and postproduction storage.[2,36,37] This microbial-generated nitrite can react with tobacco alkaloids and secondary and tertiary amines and form TSNA, volatile nitrosamines, and nitrosoacids.[8,27]

Because mycotoxins and nitrosamines are formed by microorganisms, the removal of microbes, as by pasteurization, has been shown to minimize the production of these agents.[12] Disinfection removes soil, soil metals, soil microbes, and agricultural chemicals. Subjecting tobacco to heat-treatment at 80–100°C eliminates microbes and lowers TSNA concentrations in snus.[12] The use of eBeam irradiation may be an alternative.[69] Other technologies such as microwave may also be useful.[70–72] Fire-curing and areca nut should be avoided. Technologies that eliminate microorganisms may reduce carcinogen levels and lower morbidity and mortality. Tracking the levels of carcinogens in products and making systematic adjustments to constantly drive these levels lower has been very effective by the GothiaTek standard.[12]

Certain ST products contain TSNA at levels that cause cancer. NNN and NNK are the most carcinogenic TSNA identified in tobacco products.[1,5,22,27,28] Animal studies and epidemiological findings demonstrate that NNN is a major contributor to the elevated cancer risks associated with ST.[5,6,27] In 2017, the FDA issued a proposed rule which, if finalized, would establish a limit of NNN in finished ST sold in the US. It would require that the mean level of NNN in any batch of finished ST not exceed 1.0 μg/g of tobacco (dry weight basis) at any time through the product's labeled expiration date as determined by specified product testing. FDA is reviewing the comments submitted on the purposed rule.[23]

Tools to lower toxicants in ST and thereby reduce the morbidity and mortality associated with use are now available. One certain remedy is abstinence. Although no tobacco is entirely safe, much can be done to reduce toxicity. In Sweden, systematic and sustained efforts have resulted in greatly decreased levels of potent carcinogens in snus.[11] For ST, the use of lower-alkaloid tobacco types grown in soils with lower nitrate and toxic metal content, washing and disinfecting at harvest (a common practice for other orally consumed products), remedial changes to curing and processing, omitting fire-curing and areca nut, and product refrigeration can prevent or reduce the accumulation, formation, or introduction of several potent carcinogens. During the life cycle of ST, agronomists, farmers, manufacturers, and regulators have the influence to achieve the goal of less harmful ST in the marketplace.

Disclaimer

The findings and conclusions in this chapter are those of the author and do not necessarily represent the official positions of the US Centers for Disease Control and Prevention. Use of trade names and commercial sources is for identification only and does not constitute endorsement by the US Department of Health and Human Services or the US Centers for Disease Control and Prevention.

References

1. Idris AM, Nair J, Ohshima H, et al. Unusually high-levels of carcinogenic tobacco-specific nitrosamines in Sudan snuff (toombak). *Carcinogenesis.* 1991;12:1115–1118.
2. Di Giacomo M, Paolino M, Silvestro D, Vigliotta G, Imperi F, et al. Microbial community structure and dynamics of dark fire-cured tobacco fermentation. *Appl Environ Microbiol.* 2007;73:825–837.
3. Pappas RS. Toxic elements in tobacco and in cigarette smoke: inflammation and sensitization. *Metallomics.* 2011;3:1181–1198.
4. Golia EE, Dimirkou A, Mitsios IK. Accumulation of metals on tobacco leaves (primings) grown in an agricultural area in relation to soil. *Bull Environ Contam Toxicol.* 2007;79:158–162.
5. Department of Health and Human Services. *Smokeless Tobacco and Public Health: A Global Perspective.* Bethesda, MD: Department of Health and Human Services, Centers for Disease Control and Prevention and National Institutes of Health, National Cancer Institute (NIH Publication No. 14-7983); 2014.
6. *WHO Study Group on Tobacco Product Regulation: Report on the Scientific Basis of Tobacco Product Regulation: Seventh Report of a WHO Study Group.* Geneva: World Health Organization; 2019 (Who Technical Report Series; no. 1015). License: CC BY-NC-SA 3.0 IGO.
7. Davis D, Nielsen M, eds. *Tobacco: Production, Chemistry and Technology.* Wiley-Blackwell; 1999.
8. Spiegelhalder B, Fischer S. formation of tobacco-specific nitrosamines. *Crit Rev Toxicol.* 1991;21:241.
9. Saleem S, Naz SA, Shafique M, Jabeen N, Ahsan SW. Fungal contamination in smokeless tobacco products traditionally consumed in Pakistan. *J Pakistani Med Assoc.* 2018;68: 1471–1477.
10. Zitomer N, Rybak ME, Li Z, Walters MJ, Holman MR. Determination of aflatoxin B1 in smokeless tobacco products by use of UHPLC-MS/MS. *Ag Food Chem.* 2015;63: 9131–9138.
11. Smyth EM, Kulkarni P, Claye E, et al. Smokeless tobacco products harbor diverse bacterial communities that differ across products and brands. *Appl Microbiol Biotechnol.* 2017. https://doi.org/10.1007/s00253-017-8282-9. Published April 22, 2017.
12. Rutqvist LE, Curvall M, Hassler T, Ringberger T, Wahlberg I. Swedish snus and the GothiaTek standard. *Harm Reduct J.* 2011;8:11.
13. Adamu CA, Bell PF, Mulchi C. Residual metal contrenterations in soils and leaf accumulations in tobacco a decade following farmland applications of municipal sludge. *Environ Pollut.* 1989;56:113–126.
14. Lawler TS, Stanfill SB, Zhang L, Ashley DL, Watson CH. Chemical characterization of domestic oral tobacco products: total nicotine, pH, unprotonated nicotine and tobacco-specific N-nitrosamines. *Food Chem Toxicol.* 2013;57:380–386.
15. Richter P, Hodge K, Stanfill S, Zhang L, Watson C. Surveillance of moist snuff: total nicotine, moisture, pH, un-ionized nicotine, and tobacco-specific nitrosamines. *Nicotine Tob Res.* 2008;10:1645–1652.
16. Stanfill SB, Connolly GN, Zhang L, et al. Surveillance of international oral tobacco products: total nicotine, un-ionized nicotine and tobacco-specific nitrosamines. *Tob Control.* 2011;20:e2.

17. Stanfill SB, Croucher R, Gupta P, et al. Chemical characterization of smokeless tobacco products from south asia: nicotine, unprotonated nicotine, tobacco-specific N-nitrosamines and flavor compounds. *Food Chem Toxicol.* 2018;118:626–634.
18. Stepanov I, Gupta PC, Dhumal G, et al. High levels of tobacco-specific N-nitrosamines and nicotine in Chaini Khaini, a product marketed as snus. *Tob Control.* 2015;24(e4): e271–e274.
19. Stepanov I, Jensen J, Hatsukami D, Hecht SS. New and traditional smokeless tobacco: comparison of toxicant and carcinogen levels. *Nicotine Tob Res.* 2008;10:1773–1782.
20. International Agency for Research on Cancer. *Agents classified by the IARC Monographs,* Volumes 1–106. Updated June 28, 2012. Available from: http://monographs.iarc.fr/ENG/Classification/ClassificationsAlphaOrder.pdf.
21. Stepanov I, Hecht SS, Ramakrishnan S, Gupta PS. Tobacco-specific nitrosamines in smokeless tobacco products marketed in India. *Intl J Cancer.* 2005;116:16–19.
22. India Ministry of Health and Family Welfare. *Smokeless Tobacco and Public Health in India. Chapter 13. Chemistry and Toxicology of Smokeless Tobacco.* Rajani A. Bhisey and Stephen B. Stanfill. Released at COP7, 8 June 2017, New Delhi, India. Ministry of Health and Family Welfare, Government of India.
23. Register F. Tobacco product standard for N-nitrosonornicotine level in finished smokeless tobacco products. *Food and Drug Admin.* January 23, 2017;82(13).
24. Sisson VA, Severson RF. Alkaloid composition of *Nicotiana* species. *Beitr Tabakforsch Int.* 1990;14:327–339.
25. Lisko JG, Stanfill SB, Duncan BW, Watson CH. Application of GC-MS/MS for the analysis of tobacco alkaloids in cigarette filler and various tobacco species. *Anal Chem.* 2013;85:3380–3384.
26. Furer V, Hersch M, Silvetzki N, Breuer GS, Zevin S. *Nicotiana glauca* (tree tobacco) intoxication—two cases in one family. *J Med Toxicol.* 2011;7(1):47–51.
27. International Agency for Research on Cancer. Smokeless tobacco and some tobacco-specific N-nitrosamines. In: *IARC Monographs on the Evaluation of Carcinogenic Risks to Humans.* Vol. 89. Lyon, France: World Health Organization, International Agency for Research on Cancer; 2007.
28. Idris AM, OIbrahim S, Vasstrand EN, et al. The Swedish Snus and the Sudanese Toombak: are they different? *Oral Oncol.* 1998;34:558–566.
29. Burton HR, Dye NK, Bush LP. Distribution of tobacco constituents in tobacco leaf tissue. 1. Tobacco-specific nitrosamines, nitrate, nitrite, and alkaloids. *J Agric Food Chem.* 1992;40(6):1050–1055.
30. Wang J, Yang H, Shi H, Zhou J, Bai R, Zhang M, Jin T. Nitrate and nitrite promote formation of tobacco-specific nitrosamines via nitrogen oxides intermediates during postcured storage under warm temperature. *J Chem.* 2017;2017. Article ID 6135215.
31. Halstead MM, Watson CH, Pappas RS. Electron microscopic analysis of surface inorganic substances on oral and combustible tobacco products. *J Anal Toxicol.* 2015;39: 698–701.
32. Fisher MT, Bennett CB, Hayes A, et al. Sources of and technical approaches for the abatement of tobacco specific nitrosamine formation in moist smokeless tobacco products. *Food Chem Toxicol.* 2012;50:942–948.
33. Tyx RE, Stanfill SB, Keong LM, Rivera AJ, Satten GA, Watson CH. Characterization of bacterial communities in selected smokeless tobacco products using 16S rDNA analysis. *PLoS One.* 2016;11:e0146939.
34. Wiernik A, Christakopoulos A, Johansson L, Wahlberg I. Effect of air-curing on the chemical composition of tobacco. *Recent Adv Tob Sci.* 1995;21:39–80.
35. Wahlberg I, Wiernik A, Christakopoulos A, Johansson L. Tobacco-specific nitrosamines. A multidisciplinary research area. *Agro Food Ind Hi Tech.* 1999;10:23–28.

36. Andersen RA, Fleming PD, Burton HR, Hamilton-Kemp TR, Sutton TG. Nitrosated, acylated, and oxidized pyridine alkaloids during storage of smokeless tobaccos: effects of moisture, temperature, and their interactions. *J Agric Food Chem*. 1991;39:1280–1287.
37. Han J, Yasser M, Sanad YM, et al. Bacterial populations associated with smokeless tobacco products. *Appl Environ Microbiol*. 2016;82:6273–6283.
38. Law AD, Fisher C, Jack A, Moe LA. Tobacco, microbes, and carcinogens: correlation between tobacco cure conditions, tobacco-specific nitrosamine content, and cured leaf microbial community. *Microb Ecol*. 2016;72:120–129.
39. Hempfling WP, Bokelman GH, Shulleeta M. United States Patent Application. Publication No. US 6,755,200 B1. Jan. 29, 2004.
40. Hearn BA, Ding YS, England L, et al. Chemical analysis of Alaskan iq'mik smokeless tobacco. *Nicotine Tob Res*. 2013;15:1283–1288.
41. Chen Z, Xia Z, Lei L, et al. Characteristic analysis of endophytic bacteria population in tobacco. *Acta Tabacaria Sinica*. 2014;20:102–107.
42. Corynebacterium. *Bergey's Manual of Systematics of Archaea and Bacteria*. Hoboken, NJ: Bergey's Manual Trust/John John Wiley and Sons; 2015.
43. Nishimura T, Vertes AA, Shinoda Y, Inui M, Yukawa H. Anaerobic growth of *Corynebacterium glutamicum* using nitrate as a terminal electron acceptor. *Appl Microbiol Biotechnol*. 2007;75:889–897.
44. González PJ, Correia C, Moura I, Brondino CD, Moura JJG. Bacterial nitrate reductases: molecular and biological aspects of nitrate reduction. *J Inorg Biochem*. 2006;100: 1015–1023.
45. Al-Hebshi NN, Alharbi FA, Mahri M, Chen T. Differences in the bacteriome of smokeless tobacco products with different oral carcinogenicity: compositional and predicted functional analysis. *Genes*. 2017;8:106.
46. Welch JLM, Rossetti BJ, Rieken CW, Dewhirst FE, Borisya GG. Biogeography of a human oral microbiome at the micron scale. *Proc Natl Acad Sci*. 2016;113:E791–E800.
47. Fant RV, Henningfield JE, Nelson RA, Pickworth WB. Pharmacokinetics and pharmaco-dynamics of moist snuff in humans. *Tob Control*. 1999;8:387–392.
48. Tomar SL, Henningfield JE. Review of the evidence that pH is a determinant of nicotine dosage from oral use of smokeless tobacco. *Tob Control*. 1997;6:219–225.
49. Federal Register. *Notice Regarding Requirement for Annual Submission of the Quantity of Nicotine Contained in Smokeless Tobacco Products Manufactured, Imported, or Packaged in the United States. 1999*. Washington DC: Centers for Disease Control and Prevention; Department of Health and Human Services; 1999.
50. Harwani D. The great plate count anomaly and the unculturable bacteria. *Int J Sci Res*. 2012;2:350–351.
51. 3M Corporation website. *3M Petrifilm Plates*. Available from: https://www.3m.com.
52. Omidbakhsh N, Ahmadpour F, Kenny N. How reliable are ATP bioluminescence meters in assessing decontamination of environmental surfaces in healthcare settings? *PLoS One*. 2014;9:e99951.
53. Silliker Inc. *Performance Evaluation of Various ATP Detecting Units*. 2010. Food Science Center report RPN 13922.
54. Moore G, Griffith C, Fielding L. A comparison of traditional and recently developed methods for monitoring surface hygiene within the food industry: a laboratory study. *Dairy, Food, Environ Sanit*. 2001;21:478–488.
55. Hygiena Corporation. Available from: www.hygiena.com.
56. Stanfill SB, Jia LT, Watson CH, Ashley DL. Rapid and chemically-selective quantification of nicotine in smokeless tobacco products using gas chromatography/mass spectrometry. *J Chromatogr Sci*. 2009;47(10):902–909.
57. Pappas RS, Stanfill SB, Watson CH, Ashley DL. Analysis of toxic metals in commercial moist snuff and alaskan iqmik. *J Anal Toxicol*. 2008;32:281–291.

58. Jain V, Garg A, Parascandola P, Khanwala SS, Stepanov I. Analysis of alkaloids in areca nut-containing products by liquid chromatography-tandem mass spectrometry. *J Agric Food Chem*. 2017;65:1977–1983.
59. Lisko JG, Stanfill SB, Watson CH. Quantitation of ten flavor compounds in unburned tobacco products. *Anal. Methods*. 2014;6(13):4698–4704.
60. Camas N, Arslan B, Uyanik A. Elemental analysis of tobacco leaves by FAAS and WDXRF spectroscopy. *Asian J Chem*. 2008;20(4):3135–3142.
61. Stanfill SB, Oliveira-Silva AL, Lisko J, et al. Comprehensive chemical characterization of south American nasal rapés: flavor constituents, nicotine, tobacco-specific nitrosamines and polycyclic aromatic hydrocarbons. *Food Chem Toxicol*. 2015;82:50–58.
62. Rein AJ, Seelenbinder J. *Handheld and Portable FTIR Spectrometers for the Analysis of Materials: Taking the Lab to the Sample*. American Laboratory; June 7, 2013. Available from https://www.americanlaboratory.com/914-Application-Notes.
63. Lewis RS. Potential mandated lowering of nicotine levels in cigarettes: a plant perspective. *Nicotine Tob Res*. 2018:nty022.
64. Lu J, Zhang L, Lewis RS, et al. Expression of a constitutively active nitrate reductase variant in tobacco reduces tobacco-specific nitrosamine accumulation in cured leaves and cigarette smoke. *Plant Biotechnol J*. 2016;14:1500–1510.
65. Julio E, Laporte F, Reis S, Rothan C, de Borne FD. Reducing the content of nornicotine in tobacco via targeted mutation breeding. *Mol Breed*. 2008;21:369–381.
66. Morita M, Shitana N, Sawada K, et al. Vacuolar transport of nicotine is mediated by a multidrug and toxic compound extrusion (MATE) transporter in *Nicotiana tabacum*. *Proc Natl Acad Sci*. 2009;106:2447–2452.
67. Shoji T, Hashimoto T. Biosynthesis and regulation of tobacco alkaloids. In: Wallner F, ed. *Herbaceous Plants: Cultivation Methods, Grazing and Environmental Impacts*. Hauppauge, New York: NOVA Biomedical; 2014.
68. Food Safety Modernization Act. Public law 111–353. Passed Januaury 4, 2011.
69. Pillai SD, Shayanfar S. Electron beam technology and other irradiation technology applications in the food industry. *Top Curr Chem*. 2017;375:6.
70. Chandrasekaran S, Ramanathan S, Basak T. Microwave food processing—a review. *Food Res Int*. 2013;52:243–261.
71. Brinley TA, Dock CN, Truong VD, et al. Feasibility of utilizing bioindicators for testing microbial inactivation in sweet potato purees processed with a continuous-flow microwave system. *J Food Sci*. 2007;72:235–242.
72. David JRD, Graves RH, Szemplenski T. *Handbook of Aseptic Processing and Packaging*. 2nd ed. Boca Raton, FL: CRC Press; 2013.
73. Lee S, Kim S, Jeong S, Park J. Effect of far-infrared irradiation on catechins and nitrite scavenging activity of green tea. *J Agric Food Chem*. 2006;54(2):399–403.
74. Rundlöf T, Olsson E, Wiernik A, et al. Potential nitrite scavengers as inhibitors of the formation of N-nitrosamines in solution and tobacco matrix systems. *J Agric Food Chem*. 2000;48(9):4381–4388.
75. Tola M, Kebede B. Occurrence, importance and control of mycotoxins: a review. *Cogent Food Agric*. 2016;2:1. https://doi.org/10.1080/23311932.2016.1191103.

Further reading

1. Huang J, Yang J, Duan Y, et al. Bacterial diversities on unaged and aging flue-cured tobacco leaves estimated by 16S rRNA sequence analysis. *Appl Microbiol Biotechnol*. 2010;88:553–562.
2. Wadud S, Michaelsen A, Gallagher E, et al. Bacterial and fungal community composition over time in chicken litter with high or low moisture content. *Br Poult Sci*. 2012;53:561–569.

CHAPTER NINE

Regulatory policy for smokeless tobacco

Mark J. Parascandola[1], **Wallace B. Pickworth**[2]
[1]National Cancer Institute, Bethesda, MD, United States
[2]Battelle Memorial Institute, Baltimore, MD, United States

ABBREVIATIONS

CPT Center for Tobacco Products
FCTC Framework Convention on Tobacco Control
FDA United States Food and Drug Administration
FSPTCA Family Smoking Prevention and Tobacco Control Act
MRTP Modified risk tobacco products
PREP Presumed reduced exposure products
WHO World Health Organization

Introduction

The purposes of the chapter are to apply research findings to the understanding of regulating smokeless tobacco (ST) products, to identify how products could be regulated, and to identify specific research questions that may facilitate regulation. Both international and US approaches to ST regulation are discussed by using case histories and published research findings.

At its most fundamental level, tobacco regulation seeks to diminish the harmful public health effects of an intrinsically harmful but legal substance.[1] Regulatory approaches to ST may center on three general areas: product access, product appeal, and product characteristics. Table 9.1 provides examples of policy and regulatory approaches for ST. In general, product regulation may involve limitation to access, changes in appeal, or mandated changes to the product that reduce its harm or appeal. The table lists elements of each approach that can be addressed by regulation. Access can be controlled by restriction of sales to customers of a certain age or restriction of use to certain places. Price is a key determinant of access, especially for youth and the economically disadvantaged. Appeal can be addressed by regulating flavorings, packaging, and advertising content and venue, and by warnings of health consequences, especially to the young adult market,

Smokeless Tobacco Products
ISBN: 978-0-12-818158-4
https://doi.org/10.1016/B978-0-12-818158-4.00009-1

Table 9.1 Regulatory approaches to smokeless tobacco products.

Access	Appeal	Characteristics
Price/taxation	Advertising	Nicotine content
Sales restrictions	Health campaigns	pH
Age restrictions	Counter advertisements	Format (loose or sachets)
Workplace restrictions	Packaging	Flavoring
Product ban	Health/risk warnings	Toxicant levels

who are at risk of transitioning to combustible tobacco. Finally, the characteristics—the format and components of the product itself—can be regulated. Nicotine content, flavors, and the availability of portioned "pouches" of moist snuff may influence product acceptability. Laboratory studies have shown that the pH of ST products is associated with the speed of nicotine absorption,[2–4] a key determinant in product liking, abuse liability, and addiction.[5,6]

The opportunity for tobacco control has never been stronger than it is today. In contrast to half a century ago, there is an extensive global community of tobacco control experts and advocates who have decades of experience enacting and implementing policies, programs, and interventions. And there is a strong and growing evidence base to support tobacco control policies and interventions and a robust global tobacco surveillance system. A global movement for tobacco control has developed around the World Health Organization (WHO) Framework Convention on Tobacco Control (FCTC), an international treaty aimed at addressing the global tobacco problem, now ratified by over 170 countries. And there is a consensus that the available evidence already supports implementation of strong control policies such as WHO's MPOWER (https://www.who.int/tobacco/control/en/).

However, until recently, global efforts focused primarily on cigarettes. The vast majority of scientific evidence and experience at the local and national level is aimed at cigarette smoking. Compared with cigarettes, the health effects of ST use, the characteristics of ST products, and the effectiveness of policies and interventions are far less understood, as previous reports have noted. This may seem appropriate, given the greater prevalence and well-documented global health burden of cigarette smoking. Yet the burden of ST is also substantial, and in some ways more complex, and warrants far greater attention than it has received.[7]

There are several challenges specific to ST that make the straightforward application of conventional tobacco control measures difficult.

- **Wide range of products in use:** Understanding the use and impact of ST is complicated by the diversity of products and user behaviors. Products with different characteristics are available around the world, including chewing tobacco, snus, gutkha, betel quid with tobacco, toombak, iqmik, and tobacco lozenges. Other ingredients such as areca nut in betel quid (both are carcinogenic) and slaked lime, which increases absorption of nicotine, also contribute to effects. Yet limited data are available on the properties of these products, how they are consumed, and what consumers are exposed to.
- **Variable results on health risks:** The magnitude of health risks directly associated with ST appears to differ across countries and regions, likely due in part to differences between products and use patterns. Laboratory analyses have shown widely varying levels of known carcinogens and nicotine in ST products from different regions, and epidemiologic studies in different regions have reached varying risk estimates for cancer and cardiovascular disease. Data to precisely quantify these differences and to identify the factors that drive them are lacking.
- **Dual product use:** In some contexts, ST users are likely to consume other tobacco products, including cigarettes. Separating the effects of the two categories can be challenging.
- **Evolving product market:** In the first decade of the 21st century, major tobacco manufacturers introduced a new range of ST products aimed at broad consumer appeal by the use of attractive flavors such as mint and fruit and new delivery methods such as lozenges and small pouches that eliminate the need to spit. Multinational cigarette companies Philip Morris and R.J. Reynolds came out with snus products carrying the well-known Marlboro and Camel brand names. While these products have had limited market success, Reynolds has submitted a modified risk tobacco product application to the FDA requesting approval to make harm reduction claims for Camel Snus.[8] Recently, novel nicotine delivery devices such as electronic cigarettes have further complicated the landscape. Evolving trends in the marketplace are likely to affect the use of ST and other tobacco products.
- **Limited treatment options:** Intervention strategies for ST use cessation have had mixed success. Clinical trials have shown that behavioral interventions in particular settings, such as dental offices, may increase

abstinence, although the available evidence is insufficient to support recommendations about the specific intervention components that should be applied. Trials of pharmacotherapies such as the nicotine patch and nicotine gum have shown limited impact on abstinence rates over 6 months, though there are some encouraging results for varenicline.[9]

Case studies

There is no clear standard for policies regulating and controlling ST products. Countries have implemented a diverse range of policies: banning of all products or some categories, requiring warning labels specific to ST, restricting advertising and sponsorship, and taking tax and price measures. In general, these measures remain more limited than for cigarettes. For example, taxes are lower on ST than on cigarettes in most countries.[7] The following four case studies illustrate the diversity of approaches.

The United States: applying tobacco control measures to ST

Cigarette packages began carrying a mandated health warning in 1965, and broadcast advertising was prohibited by law in 1971. These provisions did not apply to ST until 1986. The Comprehensive Smokeless Tobacco Health Education Act required placement of government-specified health warnings on packages and in advertisements and banned advertising on television and radio. This law required rotating warning labels: "Warning: This product may cause mouth cancer"; "Warning: This product may cause gum disease and tooth loss"; and "Warning: This product is not a safe alternative to cigarettes." The law also required the Federal Trade Commission to report annually to Congress on advertising and promotional activities, information that was already being collected for cigarettes.[10]

The tobacco industry's self-regulation practices were also weaker for ST than for cigarettes. For example, the cigarette companies had a voluntary code that prohibited distribution of free samples to persons under age 21, while the ST industry's code set the age at 18. This younger age cutoff allowed them to conduct promotional campaigns targeted to college students, including free product sampling on college campuses. The cigarette industry code prohibited the use of testimonials by athletes, and other celebrities perceived to appeal to the young, but ST advertising continued to feature well-known sports figures such as Walt Garrison, Tom Seaver, and Shep Messing. The Massachusetts Department of Public Health reviewed ST advertising in major youth-oriented magazines before and after the Smokeless Tobacco Master Settlement Agreement, which included the same restrictions on youth-targeted advertising as the Master Settlement Agreement signed by the cigarette manufacturers. The study showed that advertising more than doubled following the Agreement (between 1997 and 2001).[11]

Levy and colleagues[12] did a comprehensive review of population-level studies evaluating the impact of US tobacco control policies addressing ST, including studies on taxes, smoke-free air laws, media campaigns, advertising restrictions, health warnings, cessation treatment policies, and youth access policies. They found evidence supporting the effectiveness of taxes, media campaigns, health warnings, and cessation policies for reducing use. In particular, there was substantial and consistent evidence over time that use is sensitive to tax increases, with price elasticity equal to or higher than that for cigarettes and higher for youths than for older adults. Media campaigns and school educational programs were also found to reduce use, though the evidence was insufficient to identify what messages are most effective. While the United States does not mandate graphic warning labels on tobacco products, experimental studies suggest that they would be effective for ST, especially among youths. Population-level cessation treatment policies such as telephone quitlines can also be effective. The evidence on youth access policies, however, suggested limited impact. As with cigarette marketing, introduction and marketing of novel ST products has been associated with increased use, suggesting that advertising and marketing restrictions might be effective, though evidence is lacking in the United States. Interestingly, studies have consistently found that smoking restrictions are associated with less ST use, suggesting that policies directed at reducing cigarette use also reduce the use of other tobacco products. But the available evidence on this point is based largely on older studies and may not reflect the current more dynamic tobacco market and multiple product use associated with marketing of ST and e-cigarettes to cigarette smokers. Overall, while many research questions remain to identify the most effective strategies, the existing literature demonstrates that ST use is responsive to conventional tobacco control policies.[13]

Sweden: harm reduction and industry standards

The "Swedish Experience" has been cited as a sort of natural experiment in the use of ST for harm reduction.[14–16] Daily smoking in Sweden dropped from 40% in 1976 to 15% in 2002 for males and 34%–20% for females.[14]

According to the Eurobarometer survey, in 2017 Sweden had the lowest smoking prevalence in the European Union, at 7% (the next lowest was the United Kingdom at 17%).[17] Meanwhile, snus use remained high among men, at 19% (4% among women).[18] Sweden also has substantially lower tobacco-related mortality. In 2000, 10% of male deaths in Sweden could be attributed to smoking, compared with 22% in Europe.[19] Nicotine intake per capita per year for the Swedish population aged 15 or over is 3043 mg, similar to that in Denmark (3014 mg). But while almost half (1400 mg) of the nicotine intake in Sweden is from ST, nicotine from ST in Denmark is negligible at 12 mg.[20]

However, in addition to snus, there are several factors that likely played a crucial role in reducing the prevalence of cigarette smoking in Sweden since the 1950s. National trends in the growth and decline of smoking can follow markedly different trajectories. Smoking gained popularity relatively late in Sweden, and it never reached the levels of some other European countries. In the early 1960s, Sweden was one of the first countries to fund an organized tobacco control effort, including the development of cessation clinics and antismoking education programs, and also had earlier access to some nicotine replacement therapies.[21] Changes in popular culture also likely had an impact as popular portrayals of smoking shifted from accepting to negative.[22] Thus, it is difficult to determine to what extent the availability and use of snus can explain low smoking prevalence. Studies of the impact of snus use on cigarette smoking patterns in Sweden have reached divergent conclusions.[23,24]

Epidemiologic studies suggest lower oral cancer risk in Swedish snus users than in ST users in other high use countries, such as the United States and India.[25] Swedish manufacturers pasteurize their snus and adhere to the voluntary GothiaTek standard implemented in the late 1990s.[26] As a consequence, their products have lower levels of toxicants. The voluntary standard includes selection of tobacco leaf, pasteurization, and cold storage in transit and on the shelf, with a maximum temperature of 8°C. However, while the lower level of toxicants has been documented, it is not clear to what degree the lower oral cancer rate can be attributed to specific toxicant levels.

European Union: snus ban

The European Union first imposed a ban on the sale of oral tobacco products in 1992,[27] following the introduction of U.S. Smokeless Tobacco Company's Skoal Bandits to the United Kingdom and other countries.[28] Language prohibiting sale of "tobacco for oral use" was also included in Article 17 of the 2014 European Union Tobacco Products Directive, with an exemption for Sweden.[29] These restrictions do not provide a comprehensive ban on all forms of ST, as the definition of "tobacco for oral use" included in the Directive specifically excludes products that are chewed and emphasizes the use of tobacco in sachets (pouches).[30] So this partial ban is, in effect, a ban on snus and moist snuff. The prohibition of snus sales within the European Union has been challenged by the Swedish Match Company and by the Swedish Government, who argue that it was a violation of free trade principles.[31] Also, transnational companies, including British American Tobacco, Philip Morris International, and Japan Tobacco International, have lobbied against the ban, citing the Swedish Experience as evidence that snus can be an effective harm reduction agent.[28] The ban has also received criticism from some in the public health community, who argue that the partial ban should be replaced with comprehensive regulation of the toxicity of all tobacco products.[32] ST use remains very low in all EU countries except Sweden.[33]

India: Gutkha ban

In 2011, new rules introduced under the national Food Safety and Standards Regulations prohibited any harmful ingredient, including nicotine and tobacco, from being added to food. The Indian Supreme Court had ruled in 2004 that gutkha, a commonly used ST, was a "food product." The 2011 rules authorized state food commissioners to ban it. In March 2012, Madhya Pradesh became the first state to do so. By October 2013, all of India's states and union territories except Meghalaya and Lakshadweep had banned its sale.[34] While some states have been relatively successful in enforcing the ban, the tobacco industry has sought to circumvent it by selling key gutkha components, pan masala, and tobacco, in separate pouches. A 2014 study by the WHO India Country Office and the Johns Hopkins Bloomberg School of Public Health found that the state-level gutkha bans have reduced use. However, it is not clear that the policy has reduced tobacco use overall, as some users may switch to other products.[35] In India, ST use is higher than smoking prevalence; the 2016–17 Global Adult Tobacco Survey for India reported that 28.6% of the population consume tobacco in some form, 10.7% smoke, and 21.4% use ST. Use of khaini (a kind of ST) alone, the most prevalent form of tobacco, is 11%.[36]

The WHO FCTC and implementation of smokeless tobacco policies

The FCTC applies to all tobacco products. Indeed, Article 1(f) defines tobacco products as including all "products entirely or partly made of the leaf tobacco as raw material which are manufactured to be used for smoking, sucking, chewing, or snuffing."[37] However, no specific guidance has been developed on how to implement the FCTC for ST. According to a recent assessment, out of 181 parties to the Convention, 135 have included ST under the tobacco products definition in their laws. Of those 135, 112 have explicitly and 23 have implicitly defined ST products.[12,38]

Key provisions in the FCTC as applied to ST have been implemented to varying degrees in some countries and not at all in others. Almost all provisions have direct and distinct implications for ST and, to be fully implemented, will require guidance specific to ST products. For example, demand-reduction measures—such as regulation of contents, packaging and labeling, education and communication efforts, and dependence and cessation interventions—would require tailoring to ST users and the context of their use.

FCTC progress reports continue to show lagging implementation. As of January 2017, 91 parties (51%) required health warnings on ST covering at least 30% of the package, compared with 137 parties (77%) that had done so for cigarettes. Only 28 parties (16%) had policies fully compliant with the guidelines for Article 11 for ST, in comparison to 95 parties for cigarettes (53%). The lack of specific guidance has likely contributed to this disparity. Also, because ST prevalence is lower than cigarette prevalence in most countries, health leaders may view it as a lower priority.[38]

Health promotion efforts and media campaigns

As with other policy and intervention areas, relatively few educational and media campaigns have been aimed at ST users compared with cigarette smokers. Therefore, there is limited evidence and experience about the effectiveness of such campaigns. However, recent national media campaigns in India and the United States were targeted at ST users, and the results have undergone rigorous evaluation. Following are highlights of those campaigns.

A mass media campaign about smokeless tobacco in India

In 2009, the government of India conducted the first mass media campaign in India highlighting the harmful effects of ST use. They developed and pretested a 30-second documentary style public service announcement featuring an oral cancer surgeon from the Tata Memorial Hospital in Mumbai describing the suffering and disfigurement seen in his oral cancer patients. The campaign had a media spend of approximately US$1 million and was aired for 6 weeks from November to December 2009. Fig. 9.1 is a poster from the campaign. A nationally representative household survey of ST users (N = 2898) was conducted to evaluate the impact.

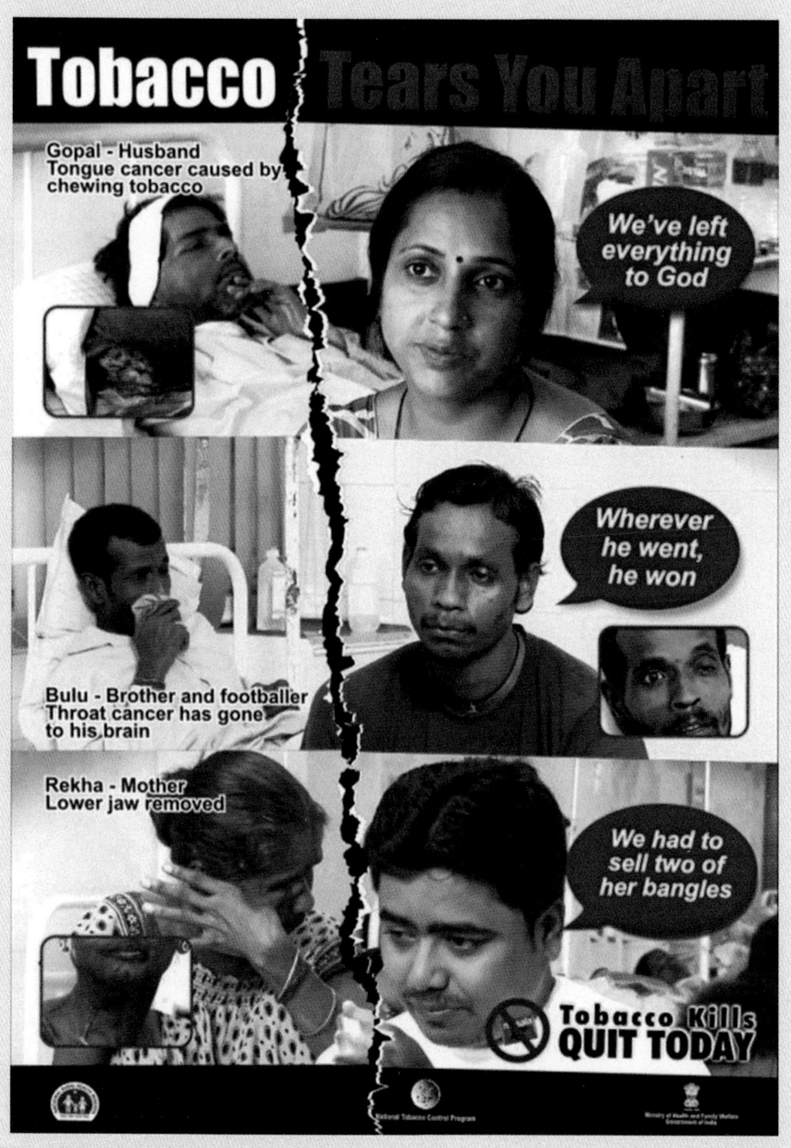

Figure 9.1 Poster from the government of India mass media campaign on smokeless tobacco.

(Continued)

A mass media campaign about smokeless tobacco in India (*cont'd*)

The campaign had the intended effect: 63% of ST-only users and 72% of dual users (those who consumed both smoking and smokeless forms) recalled the advertisement, primarily through television delivery. The majority (over 70%) of those aware of the campaign said that it made them stop and think, was relevant to their lives, and provided new information. Seventy-five percent of ST-only users and 77% of dual users said that it made them feel concerned about their habit. Campaign awareness was associated with better knowledge about the cancer effects of ST use, more negative attitudes toward ST, and greater cessation-oriented intentions and behaviors among users. Notably, campaign awareness was also associated with cessation-oriented behaviors: aware smokeless-only users were significantly more likely than unaware users to seriously consider quitting, to have stronger beliefs in their own ability to quit, and to have tried to quit in the previous 2 months. While the evaluation was a cross-sectional survey and could not measure actual cessation outcomes, these cessation-related behaviors are encouraging.[62] Subsequent analysis found the campaign to be highly cost-effective as well.[63]

The FDA "Real Cost" campaign about smokeless tobacco

The United States Food and Drug Administration launched its first tobacco prevention campaign, "The Real Cost," in 2014 to educate at-risk teens on the harmful effects of cigarette smoking. In 2016, the campaign expanded to educate rural boys, a high risk group, on the harms of ST use. Many youths at risk are not aware of the negative health consequences of ST use. The ST campaign targeted rural, white, male teenagers about risks of dipping: nicotine addiction, gum disease, tooth loss, and multiple kinds of cancer. The central message was that "smokeless doesn't mean harmless." Advertising was placed in 35 US markets specifically selected to reach the target audience (https://www.fda.gov/TobaccoProducts/PublicHealthEducation/PublicEducationCampaigns/TheRealCostCampaign/default.htm#TobaccoPreventionCampaign).

The results have not yet been evaluated, but the initial campaign focused on youth smoking was found to be highly impactful and cost-effective. It was estimated to have prevented 348,398 youths aged 11–18 years from initiating smoking during the period February 2014 to March 2016 (https://www.cdc.gov/mmwr/volumes/66/wr/mm6602a2.htm).

The economics of smokeless tobacco use

Tobacco use patterns also exhibit substantial economic disparities. Use is consistently higher among lower-income and lower-education segments of the population, a pattern which occurs in high-income countries as well.[39] When use is concentrated in lower-income groups, it may have indirect effects on family finances, health, and education beyond the direct health effects. For example, studies on the nexuses among tobacco use, nutrition, human capital investments, and poverty demonstrate that expenditures on smoking constitute a significant portion of household budget, which can lead to reduced spending on basic commodities like food, health, education, housing, transport, and energy. This phenomenon is known as the crowding-out effect, which in turn may exacerbate the effects of poverty, including the impact on nutritional status of children. The extant literature has focused primarily on smoking. However, given the high prevalence of ST use in very poor populations, it would be expected to have a similar impact. Indeed, using household economic data from a national survey in Bangladesh, where ST is highly prevalent, Husain and colleagues found that tobacco user households on average allocated less in clothing, housing, education, energy, transportation, and communication than nonuser households.[40]

Product regulation

The WHO Study Group on Tobacco Product Regulation has addressed research and regulatory needs related to ST products in two recent reports. A 2008 publication urged that all of them be subjected to comprehensive regulatory control by an independent scientific government agency. Moreover, the study group noted that "research on the health hazards and risks to individuals and populations of use of ST products is essential for governments and for implementation of the Framework Convention [on Tobacco Control]." This research should address how the design and manufacture of tobacco products could be modified to alter their health effects.[41] A subsequent report in 2010 proposed establishing upper limits for two tobacco-specific nitrosamines (TSNA), N-nitrosonornicotine (NNN) and 4-(methylnitrosamino)-1-(3-pyridyl)-1-butanone (NNK), and one polycyclic aromatic hydrocarbon, benzo(a)pyrene (BaP), in ST products. This report also recommended that regulation of the distribution and sale of ST should include measures to limit increases in TSNA, including storage requirements and expiration dates.[42] In 2017, the FDA Center for Tobacco

Products (CTP) proposed a rule limiting the content of NNN, a known carcinogen. Although the authors acknowledged that existing evidence is not sufficient to establish whether reducing the levels of individual constituents will have a measurable impact on cancer risks, they asserted that "it is difficult to justify allowing high levels of known carcinogenic constituents in a product that is known to cause cancer, when lower levels are readily achievable with existing technology."[43] In 2017, the CTP proposed a rule setting the upper limit of NNN to 1 μg per gram of tobacco (dry weight) in finished products.

Harm reduction

The response to the hazards of ST use is complicated by discussions about the possibility of use as a means of harm reduction for smokers.[44] Some scientists have suggested that ST may reduce harm by providing an alternative to cigarettes; that is, smokers who switch completely to ST, which does not carry the same risk of lung cancer and respiratory disease, might reduce the overall risk.[14] While ST also causes cancer and other diseases, the overall risk for a lifetime ST user would be expected to be lower than for a lifetime cigarette smoker.[45] This inference requires several assumptions, however. Given the tremendous diversity of ST products and patterns of use around the world, it is difficult to support broad generalizations about the level of harm associated with ST as a category. Little is known about the constituents of some products or the amount of exposure users receive from them. Also, it is essential to consider the overall population-level impact of increased ST use. For example, will increased promotion of ST products lead to an increase in tobacco use initiation, or have an adverse impact on smoking cessation efforts?[46] Although the body of evidence is expanding, definitive studies are lacking. In short, there remain more questions than answers. Discussions of harm reduction have been limited primarily to high-income countries in North America and Western Europe, where cigarette smoking is the predominant form of tobacco use and there is a long history of tobacco control measures, and therefore may lack relevance for low- and middle-income countries.[47]

Regulation in the United States

For decades, the tobacco industry had unfettered access to the marketplace. They bolstered sales and increased product use through aggressive marketing campaigns, advertisement, unsubstantiated health claims, and appeal to youth. Since the demonstration of tobacco-related harm, the immense public health burden associated with tobacco has led to an outcry from the public health community, affected groups, individuals, states, and insurance providers that bear the economic and emotional costs. This led to the passage of the Family Smoking Prevention and Tobacco Control Act (FSPTCA) of 2009.[48] This Congressional legislation was the first meaningful attempt to place under one regulatory agency, the FDA, control of the content, advertising, and marketing of tobacco products. The original legislation specified cigarettes, moist ST, and roll-your-own tobacco and was extended to all tobacco and tobacco-derived products, including cigars, cigarillos, water pipe tobacco, and electronic cigarettes, in 2017. Although the provisions of the act grant FDA a wide regulatory authority, the agency is not allowed to mandate the reduction of nicotine to zero, nor can it change taxation (ultimately, price).

The family smoking prevention and tobacco control act

Decades of laboratory, social, and behavioral research on the immense public health consequences of tobacco use preceded congressional passage of the FSPCTA in 2009 (H.R. 1256). This law extended regulatory authority over tobacco to the FDA, empowering it to establish the Center for Tobacco Products.

Key components of the legislation relevant to ST (adapted language from FSPTCA):

1. ST, cigarettes, and roll-your-own tobacco are subject to the provisions.
2. The FDA has regulatory authority over the labeling, manufacture, sale, and contents of tobacco products—but no authority over regulations associated with agriculture and taxation.
3. ST is defined as a tobacco product.
4. Contents of ST products are subject to reporting and regulation.
5. Nicotine content of tobacco products can be regulated and reduced, but not eliminated.
6. Flavors and other additives can be regulated.
7. All ST packages must contain one of the following health warning labels:
    WARNING: This Product Can Cause Mouth Cancer
    WARNING: This Product can Cause Gum Disease and Tooth Loss
    WARNING: This Product is Not A Safe Alternative to Cigarettes
    WARNING: Smokeless Tobacco is Addictive
8. Where and how ST products are sold can be regulated.

As described below, policies that increase price to the consumer—notably taxes—reduce demand. This has been empirically shown: a 10% increase in cigarette price leads to a 3% reduction in sales. Studies using experimental and hypothetical purchasing tasks have shown that the cost of a product influences demand. Tobacco price sensitivity is most readily observed in young users and lower socioeconomic status groups. Although most of the data have been derived from cigarette use, some has demonstrated similar results in oral tobacco use.

Levy and colleagues[12] reviewed the effectiveness of tobacco control policies on ST use in the United States (NTR 2018). They reported that taxes (price), media campaigns, health warnings, and cessation treatment programs reduced ST use, but policies that limited youth access were less effective. They concluded that policies effective for cigarette regulation (price, health warnings, access, advertising, and marketing) were generally effective for ST.

Hawkins and colleagues (2018)[49] reported on the impact of tobacco control policies on adolescent ST and cigar use. Using a difference in difference regression model on data from the Youth Risk Behavior Surveillance System from 36 states between 1999 and 2013, they found that neither chewing tobacco nor cigar taxes had an effect on ST but a 10% increase in cigarette taxes increased ST use in males by 1%, cigar use in males by 1.5%, and cigar use in females by 0.7%. The authors noted that federal cigarette taxes averaged over $1 per pack but ST taxes were $0.03. Others have shown that increased cigarette taxes have led to increased use of roll-your-own cigarettes.[50] The findings lead to the conclusion that taxation of one product may cause increased use of a cheaper tobacco alternative and emphasize the need for a comprehensive taxation approach.

Similarly, with enforcement of the ban on cigarette flavors (except menthol) that occurred with the passage of the FSPTCA in 2009, some adolescents have turned to cigars and ST. There is data to suggest that the flavor of ST is an important determinant of use, particularly among youths.[49] Recently the FDA suggested that it will ban flavoring of certain tobacco products and electronic cigarettes to reduce their attractiveness to youth.[51] The calculation of the overall public health benefit of such regulations must include consideration of the unintended consequences of increasing demand, sales, and use of other less regulated products.

Prohibition of indoor and workplace smoking has been shown to increase ST use.[52,53] Many smokers use ST in places where smoking is prohibited.[12,54] Workplace and public smoking bans that were implemented to decrease exposure to nonsmoking colleagues may have led to an increase in ST use. A marketing strategy of ST sellers is to advertise their products to smokers for use in an environment where smoking is prohibited. The anti-spitting laws that sought to reduce the spread of tuberculosis led to reduced oral tobacco use, but moist snuff and snus are "spitless" (users swallow the juices).[55]

Dual use of cigarettes and smokeless tobacco

ST functions as a modified risk tobacco product (MRTP) for smoking cessation. The FSPTCA criteria for MRTP evaluation using a public health standard specifically address a need to evaluate MRTP use concurrent with continued use of an existing harmful form of tobacco such as cigarettes, with the goal of switching entirely to the less harmful.

The transition between cigarettes and ST was examined by using two lateral cohorts in the United States: exclusive smokers, exclusive ST users, dual users, and nonusers from samples in 2002–03 and 2010–11.[54] Cigarette quit rates increased from 11% to 24% among male smokers, but the corresponding ST quit rates were stable at 41% and 40%. Among women, similar results were found in exclusive cigarette users, and the findings were consistent across socioeconomic groups. The data indicate that smoking cessation is increasing but ST use has remained stable. Messer and colleagues[56] reported that dual users were more likely to make a cigarette cessation attempt than exclusive cigarette users, but they were less likely to achieve smoking abstinence. These two studies illustrate the stability of ST use over the past decade, and that it is likely fostered by the mistaken belief that it will facilitate smoking cessation. These data further illustrate unintended consequences of policy and societal changes when one product (cigarettes) is regulated and alternative products are relatively ignored.

Modified risk tobacco products

Until a product is designated an MRTP by the FDA, it cannot be introduced into interstate commerce for that purpose. The bar for acceptance is high. The following text box summarizes scientific criteria that must be addressed in the application.[48]

Generally, the applicant must present scientific evidence that the product has a health benefit by reducing the risk of tobacco-related diseases to the individual user and will benefit the health of the population at large. In practice, this has been very difficult. To date, three applications have been

FDA requirements for modified risk tobacco products

The FSPTCA (Section 911) has provisions for the marketing of tobacco products that have less risk than conventional smoking. In older tobacco literature, they were called potential reduced exposure products (PREPs). The name is now MRTP. To gain this designation, the applicant must address the following criteria:

1. The product must be approved by the FDA CTP as an MRTP.
2. The applicant must demonstrate that the product will significantly reduce harm and the risk of tobacco-related disease to the individual user and benefit the health of the population, taking into account both users of tobacco products and persons who do not currently use tobacco products.
3. The applicant must demonstrate that the overall reduction in exposure to the harmful substances is substantial, that the substances are harmful, and that, as actually used, there is a specified reduction in exposure.
4. The product will not expose consumers to higher levels of other harmful substances.
5. There will be extensive requirements for postmarketing surveillance.

submitted. Swedish Match applied for 10 (later reduced to 8) General Snus products similar to those used in Norway and Sweden.[57] British American Tobacco and its subsidiary Reynolds American applied for six styles of Camel Snus.[58] U.S. Smokeless Tobacco applied for Copenhagen Fine Cut.[59] Two applications are still in the review process (https://www.fda.gov/tobacco-products/advertising-and-promotion/modified-risk-tobacco) but the Swedish Match General Snus products were recently approved.

Knowledge gaps and recommendations

As discussed in this and preceding chapters, knowledge gaps in ST research need to be addressed to inform regulation in the domestic and international markets. Some key challenges:

- In many regions, even where ST use is highly prevalent, policies and programs for prevention and cessation are generally weaker than those that address smoked tobacco. Prices are lower, warning labels are weaker, surveillance is less developed, fewer proven interventions are available, and fewer resources are devoted to control.
- ST poses substantial challenges to regulatory efforts because of the wide variety of products and production methods. Regulation is problematic where products are not standardized because they are manufactured in

local cottage industries. An estimated 91.3% by volume of ST products worldwide are made and sold that way.[60] These products, widely used in countries such as India and Bangladesh, are assembled by local vendors and often tailored to the customer's preference. Therefore, the ingredients and other characteristics, including levels of nicotine and toxic constituents, can vary widely from one sample to another.

- ST products often contain ingredients other than tobacco, including flavorings and other additives. Areca nut, a standard ingredient in betel quid and ST products widely used in South Asia, is itself carcinogenic and dependence producing. Betel quid and areca nut are consumed by hundreds of millions of people, with or without tobacco. Areca nut is not directly addressed by tobacco control policies.[61]

Changing tobacco industry marketing strategies may influence the future public health impact of ST. In some high-income countries where restrictions on public smoking have increased and smoking prevalence has decreased, tobacco companies have marketed oral tobacco to smokers. The impact of this trend in smoking behavior and dual or poly-tobacco use remains uncertain. At the same time, multinational tobacco companies have an increasing presence among low- and middle-income countries with both smoked and smokeless products.

Several research areas related to ST warrant ongoing attention to inform future control and regulation efforts, including the following: further characterization of diverse products and ingredients, better understanding of the health effects associated with different product types and practices, how existing evidence-based control measures can be most effectively adapted to address smoking, and novel dependence treatment interventions tailored to ST users.

Conclusion

In some parts of the world ST use remains low compared with cigarette smoking, whereas in others it continues to have a substantial impact. The diversity of product characteristics, user behaviors, and regulatory environments creates obstacles for a universal approach to research and regulation. In low- and middle-income countries, where most use is concentrated, fewer resources are available. In some high-income countries, the marketing of novel products by major tobacco manufacturers creates new challenges. Nevertheless, some product standards, such as upper limits for TSNA, have been proposed. To what degree such standards can affect public health remains uncertain. The path toward regulation and control of ST use around the globe will require further research and continued effort to adapt and implement existing evidence-based interventions.

References

1. World Health Organization (WHO). *The Scientific Basis of Tobacco Product Regulation.* Geneva, Switzerland: WHO Press; 2007.
2. Fant RV, Henningfield JE, Nelson RA, et al. Pharmacokinetics and pharmacodynamics of moist snuff in humans. *Tob Control.* 1999;8(4):387–392.
3. Pickworth WB, Rosenberry ZR, Gold W, et al. Nicotine absorption from smokeless tobacco modified to adjust pH. *J Addict Res Ther.* 2014;5(3):1000184.
4. Ayo-Yusuf OA, Swart TJ, Pickworth WB. Nicotine delivery capabilities of smokeless tobacco products and implications for control of tobacco dependence in South Africa. *Tob Control.* 2004;13(2):186–189.
5. Fant RV, Pickworth WB, Henningfield JE. The addictive effects of nicotine are related to the speed of delivery. In: Opitz K, ed. *Nicotine as a Therapeutic Agent, Immunity, and the Environment.* Stuttgart: Gustav-Fisher; 1997:53–61.
6. Henningfield JE, Fant RV. Tobacco use as drug addiction: the scientific foundation. *Nicotine Tob Res.* 1999;1(suppl 2):S31–S35.
7. U.S. NCI and WHO. *The Economics of Tobacco and Tobacco Control.* NCI Tobacco Control Monograph 21. NIH Publication No. 16-CA-8029A. Bethesda: U.S. DHSS, NIH, NCI, and Geneva: WHO; 2016.
8. https://www.fda.gov/downloads/AdvisoryCommittees/CommitteesMeetingMaterials/Tobacco ProductsScientificAdvisoryCommittee/UCM620063.pdf.
9. Ebbert JO, Elrashidi MY, Stead LF. Interventions for smokeless tobacco use cessation. *Cochrane Database Syst Rev.* October 26, 2015;(10):CD004306.
10. NIH, NCI. *Smokeless Tobacco or Health: An International Perspective.* Smoking and Tobacco Control Monograph 2; September 1992. http://cancercontrol.cancer.gov/tcrb/monographs/2/m2_complete.pdf.
11. National Cancer Institute. *The Role of the Media in Promoting and Reducing Tobacco Use.* Tobacco Control Monograph No. 19. NIH Publication No. 07-6242; August 2008.
12. Levy DT, Mays D, Boyles RG, Tam J, Chaloupka FJ. The effect of tobacco control policies on US smokeless tobacco use: a structured review. *Nicotine Tob Res.* 2018; 20(1):3–11.
13. The effect of tobacco control policies on US smokeless tobacco use: a structured review D.T. Levy, D. Mays, R.G. Boyle, J. Tam, F.J. Chaloupka.
14. Foulds J, Ramstrom L, Burke M, Fagerstrom K. Effect of smokeless tobacco (snus) on smoking and public health in Sweden. *Tob Control.* 2003;12(4):349–359.
15. Ramström LM, Foulds J. Role of snus in initiation and cessation of tobacco smoking in Sweden. *Tob Control.* 2006;15:210–214.
16. Roth HD, Roth AB. Health risks of smoking compared to Swedish snus. *Inhal Toxicol.* 2005;17:1–8.
17. https://ec.europa.eu/health/tobacco/eurobarometers_en.
18. Public Health Agency of Sweden The National Survey of Public Health: Tobacco. Available online: https://www.folkhalsomyndigheten.se/folkhalsorapportering-statistik/statistikdatabaser-och-visualisering/nationella-folkhalsoenkaten/resultat-a-o/[Ref list].
19. Peto R, Lopez AD, Boreham J, Thun M, Heath C. *Mortality from Smoking in Developed Countries, 1950–2000: Indirect Estimates from National Vital Statistics.* Oxford: Oxford University Press; 1994. Available at: http://rum.ctsu.ox.ac.uk/~tobacco/.

20. Fagerstrom K. The nicotine market: an attempt to estimate the nicotine intake from various sources and the total nicotine consumption in some countries. *Nicotine Tob Res.* 2005;7(3):343–350.
21. Mitchell K, Wellings K. *Improving Health Status in Europe: A Case Study Approach to the Identification of Best Practice.* Best Practice in Smoking: the Swedish Case Study. London School of Hygiene and Tropical Medicine; September 1998.
22. Torell U. *Den Rökande Människan: Bilden Av Tobaksbruk I Sverige Mellan 1950- Och 1990-tal.* Stockholm, Sweden: Carlsson bokförlag; 2002.
23. Ramström L, Borland R, Wikmans T. Patterns of smoking and snus use in Sweden: implications for public health. *Int J Environ Res Public Health.* November 9, 2016; 13(11):E1110.
24. Tomar SL, Connolly GN, Wilkenfeld J, Henningfield JE. Declining smoking in Sweden: is Swedish Match getting the credit for Swedish tobacco control's efforts? *Tob Control.* 2003;12:368–371.
25. NCI and CDC. *Smokeless Tobacco and Public Health: A Global Perspective.* Bethesda, MD: HHS, CDC, NIH, NCI, NIH Publication No. 14-7983; December 2014. http://cancercontrol.cancer.gov/brp/tcrb/global-perspective/index.html.
26. Rutqvist LE, Curvall M, Hassler T, Ringberger T, Wahlberg I. Swedish snus and the GothiaTek® standard. *Harm Reduct J.* May 16, 2011;8:11.
27. European Commission (1992) Council Directive 92/41/EEC of 15 May 1992 Amending Directive 89/622/EEC on the Approximation of the Laws, Regulations and Administrative Provisions of the Member States Concerning the Labelling of Tobacco Products. Official Journal of the European Communities L158: 30–33.
28. Peeters S, Gilmore AB. Transnational tobacco company interests in smokeless tobacco in Europe: analysis of internal industry documents and contemporary industry materials. *PLoS Med.* 2013;10(9):e1001506. http://journals.plos.org/plosmedicine/article?id=10.1371/journal.pmed.1001506.
29. European Union. Directive 2014/40/EU of the European Parliament and of the Council of 3 April 2014 on the approximation of the laws, regulations and administrative provisions of the Member States concerning the manufacture, presentation and sale of tobacco. *Official J Europ Commun, 2014 L 127/1.* 29/04/2014.
30. https://ec.europa.eu/health/sites/health/files/tobacco/docs/dir_201440_en.pdf.
31. Snus: EU Ban on Snus Sales. Tobacco Tactics. http://www.tobaccotactics.org/index.php/Snus:_EU_Ban_on_Snus_Sales.
32. Bates C, Fagerström K, Jarvis MJ, Kunze M, McNeill A, Ramström L. European Union policy on smokeless tobacco: a statement in favour of evidence based regulation for public health. *Tob Control.* 2003;12(4):360–367.
33. Leon ME, Lugo A, Boffetta P, et al. Smokeless tobacco use in Sweden and other 17 European countries. *Eur J Public Health.* October 2016;26(5):817–821.
34. *Smokeless Tobacco and Public Health in India.* New Delhi: Ministry of Health and Family Welfare, Government of India; 2017. Ministry of Health and Family Welfare, Government of India.
35. http://www.searo.who.int/india/mediacentre/releases/2014/gutka_study/en/.
36. Tata Institute of Social Sciences (TISS). *Mumbai and Ministry of Health and Family Welfare, Government of India. Global Adult Tobacco Survey GATS 2 India 2016-17.* Ministry of Health and Family Welfare; 2018. https://mohfw.gov.in/newshighlights/global-adult-tobacco-survey-2-gats-2-india-2016-17-report.
37. https://www.who.int/tobacco/framework/WHO_FCTC_english.pdf.
38. Mehrotra R, Sinha D, Szilagyi T. Global smokeless tobacco control policies and their implementation. *Executive Summary.* 2018. https://doi.org/10.13140/RG.2.2.31229.05608.
39. https://www.who.int/tobacco/publications/economics/9789241507820/en/.

40. Husain MJ, Datta BK, Virk-Baker MK, Parascandola M, Khondker BH. The crowding-out effect of tobacco expenditure on household spending patterns in Bangladesh. *PLoS One*. 2018;13(10):e0205120.
41. https://www.who.int/tobacco/global_interaction/tobreg/publications/tsr_951/en/.
42. https://www.who.int/tobacco/global_interaction/tobreg/publications/tsr_955/en/.
43. Food and Drug Administration (FDA). *Tobacco Product Standard for N-Nitrosonornicotine Level in Finished Smokeless Tobacco Products*; 2017. https://www.federalregister.gov/documents/2017/01/23/2017-01030/tobacco-product-standard-for-n-nitrosonornicotine-level-in-finished-smokeless-tobacco-products.
44. Hatsukami DK, Lemmonds C, Tomar SL. Smokeless tobacco use: harm reduction or induction approach? *Prev Med*. 2004;38(3):309–317.
45. Levy DT, Mumford EA, Cummings KM, et al. The relative risks of a low-nitrosamine smokeless tobacco product compared with smoking cigarettes: estimates of a panel of experts. *Cancer Epidemiol Biomark Prev*. 2004;13:2035–2042.
46. Mejia AB, Ling PM, Glantz SA. Quantifying the effects of promoting smokeless tobacco as a harm reduction strategy in the USA. *Tob Control*. 2010;19(4):297–305.
47. Ayo-Yusuf OA, Burns DM. The complexity of 'harm reduction' with smokeless tobacco as an approach to tobacco control in low-income and middle-income countries. *Tob Control*. 2012;21(2):245–251.
48. *Family Smoking Prevention and Tobacco Control Act—An Overview*; 2009. https://www.fda.gov/TobaccoProducts/GuidanceComplianceRegulatoryInformation/ucm246129.htm.
49. Hawkins SS, Bach N, Baum CF. Impact of tobacco control policies on adolescent smokeless tobacco and cigar use: a difference-in-differences approach. *BMC Public Health*. 2018;18(1):154.
50. Sureda X, Fu M, Martínez-Sánchez JM, et al. Manufactured and roll-your-own cigarettes: a changing pattern of smoking in Barcelona, Spain. *Environ Res*. 2017;155:167–174.
51. *FDA Statement on Proposed Steps to Protect Youth by Preventing Access to Flavored Tobacco Products and Banning Menthol in Cigarettes*; 2018. https://www.fda.gov/NewsEvents/Newsroom/PressAnnouncements/ucm625884.htm.
52. Tauras JA, Peck RM, Cheng KW, Chaloupka FJ. Graphic warning labels and the cost savings from reduced smoking among pregnant women. *Int J Environ Res Public Health*. 2017;14(2).
53. Grossman M, Chaloupka FJ. Cigarette taxes. The straw to break the camel's back. *Public Health Rep*. 1997;112(4):290–297.
54. Chang JT, Levy DT, Meza R. Examining the transitions between cigarette and smokeless tobacco product use in the United States using the 202-2003 and 2010–2011 longitudinal cohorts. *Nicotine Tob Res*. 2018;20(11):1412–1416.
55. Toth A, Fackelmann J, Pigott W, Tolomeo O. Tuberculosis prevention and treatment. *Can Nurse*. 2004;100(9):27–30.
56. Messer K, Vijayaraghavan M, White MM, et al. Cigarette smoking cessation attempts among current US smokers who also use tobacco. *Addict Behav*. 2015;51:113–119.
57. Swedish Match North America Inc., MRTP https://www.fda.gov/TobaccoProducts/Labeling/MarketingandAdvertising/ucm533454.htm.
58. Camel Snus MRTP Application https://www.fda.gov/downloads/AdvisoryCommittees/CommitteesMeetingMaterials/TobaccoProductsScientificAdvisoryCommittee/UCM620817.pdf.
59. U.S. Smokeless Tobacco Company Modified Risk Tobacco Products (MRTP) Application https://www.fda.gov/TobaccoProducts/Labeling/MarketingandAdvertising/ucm619683.htm.

60. National Cancer Institute/Centers for Disease Control and Prevention. *Smokeless Tobacco and Public Health: A Global Perspective*. Bethesda, MD: HHS, CDC, NIH, NCI, NIH Publication No. 14-7983; December 2014. http://cancercontrol.cancer.gov/brp/tcrb/global-perspective/index.html.
61. Mehrtash H, Duncan K, Parascandola M, et al. Defining a global research and policy agenda for betel quid and areca nut. *Lancet Oncol*. December 2017;18(12):e767–e775.
62. Murukutla N, Turk T, Prasad VC, et al. Results of a national mass media campaign in India to warn against the dangers of smokeless tobacco consumption. *Tob Control*. 2012; 21(1).
63. Murukutla N, Yan H, Wang S, et al. Cost-effectiveness of a smokeless tobacco control mass media campaign in India. *Tob Control*. 2018;27(5):547–551.

Index

'*Note*: Page numbers followed by "f" indicate figures, "t" indicates tables and "b" indicates boxes.'

U

V

W

Printed in the United States
By Bookmasters